Textbooks in Electrical and Electronic Engineering

Series Editors
## G. Lancaster   E. W. Williams

# The CD-ROM and Optical Disc Recording Systems

■

## E. W. Williams

*Professor of Electronic Engineering*
*Electronic Engineering Group*
*Keele University*

OXFORD   NEW YORK   TOKYO
OXFORD UNIVERSITY PRESS
1994

Oxford University Press, Walton Street, Oxford OX2 6DP

Oxford   New York   Toronto
Delhi   Bombay   Calcutta   Madras   Karachi
Kuala Lumpur   Singapore   Hong Kong   Tokyo
Nairobi   Dar es Salaam   Cape Town
Melbourne   Auckland   Madrid

and associated companies in
Berlin   Ibadan

Oxford is a trade mark of Oxford University Press

Published in the United States
by Oxford University Press Inc., New York

© E. W. Williams, 1994

A catalogue record for this book is available from the British Library

Library of Congress Cataloging in Publication Data
Williams, E. W.
The CD-ROM and optical disc recording systems / E. W. Williams.
(Textbooks in electrical and electronic engineering; 2)
Includes index.
1.   CD-ROM.   I.   Title.   II.   Series.
TK7895.M4W55      1994      621.39'767—dc20      94–1194

ISBN 0 19 859373 2 (hbk)

Typeset by EXPO Holdings, Malaysia
Printed in Great Britain by
Biddles Ltd, Guildford & King's Lynn

# Preface

This book was originally designed to introduce CD-ROM and other optical recording systems to undergraduates and postgraduates in science and engineering. The brief for the book was widened since the growth of CD-ROM from 1990 onwards has been exponential. Multimedia systems in which audio, video, and data recording are all combined are being set up in business training centres, schools, and academic institutions. Optical recording is the heart of multimedia. For multimedia, erasable optical recording is essential and this book clearly describes how the two competing systems work—the magneto-optic and the phase change type.

All those who use computer systems in their studies or in their businesses should find it useful. It will help them to understand an erasable technology that will eventually replace the floppy disc and CD-ROM, the publishing media of the future, will become as common as CD-audio.

This book could not have been written without the help of information provided by a large number of people and companies. Information from Philips, Digital, Sony, and seventeen Japanese companies supplied during and after a DTI mission in the summer of 1992 was invaluable. Chris Coughlan, Chris Hall, and Kate Hill were members of the mission. David Williams of Williams Associates provided the latest details of optical drives and Ray Gilson filled in the gaps in my knowledge.

Brian Bowler, who tragically passed away before the book was published, prepared all the diagrams in the first draft.

Finally, the book is dedicated to my wife Margaret. She typed the first draft and improved the English in the process and put up with my absence for the many hours of research that went into this book.

*Keele*                                                                         E.W.W.
September 1993

# Contents

# Introduction to optical recording

## 1.1 The basic optical recording system

Optical recording systems in general consist of:

- media;
- optical Head (source and detector);
- media drive/system;
- interface to computer system;
- computer system for control and data handling.

The media in the form of a **hard disc** stores the data, and for commercial optical systems the capacity on a 130 mm diameter disc varies from 300 MBytes to 500 MBytes per side.

An **optical head** has a laser light source, an optical system to focus the light on to the media disc recording layer, and a detector to convert the light returning from the disc to an electronic signal.

The function of the **drive mechanism** is to take that electronic signal, amplify it and carry out signal processing on it. The drive also powers the laser and supplies the error correction code, data signals and error correction. In addition, all the controls for the optical head and the disc media motor are derived from the drive.

The drive is connected to an electronic **interface** which is compatible with a wide range of computer systems.

**Computer systems** implement control functions and supply the data which comes either from a human user, pre-recorded media, a telephone exchange, a mainframe computer system, an instrument, or a satellite.

## 1.2 Optical recording media

There are three types of optical disc in use:

- CD-ROM
- WORM
- WREM or Erasable

**Fig. 1.1** Recording media in the form of a magneto-optic disc produced by Philips/ DuPont containing 1 GBytes of data.

**Fig. 1.2** Compact disc produced by Digital showing a CD-ROM (read only memory) reading head. Note the manufacturer's printing over the working surface of the disc.

## CD-ROM stands for compact disc read-only memory

The CD-ROM media is made up of a series of pits in the surface of a polycarbonate substrate (see Fig. 1.3(a)). These pits are varied in length in order to code the data. In order to read the data a reflective thin film is coated on to the pit surface of the subtrate. When light is focussed on to the pit it is scattered by the edges of the pit. Light reflected from the 'land' between the pits is scattered very little, so that a much stronger signal may be received by a detector system. The high and low reflectivity signals then enable the digital coding to be read effectively by the detector.

Once the CD-ROM disc has been fabricated the data cannot be altered, and this is why it is called a read-only memory.

## WORM stands for write-once read-many times

As Fig. 1.3(b) shows, the WORM has a special reflective layer which can be melted by the laser to form a pit or a hole. The laser operates in the high power mode to melt the layer. The reflectivity difference between the reflective layer and the hole can be used to read the data in a similar way to the CD-ROM.

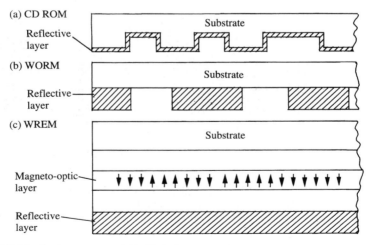

**Fig. 1.3** The basic structure of (a) CD-ROM, (b) WORM and (c) WREM optical discs (cross-sectional views). The light source writes/reads from above.

Unlike the CD-ROM, however, data can be written on the disc by the computer operator. However, this data cannot be rewritten and cannot be erased.

WREM stands for write-read-erase memory

This type of memory is also called erasable and rewritable. It is equivalent to the magnetic Winchester disc in that the memory can be written, erased, rewritten and erased as many times as the computer operator desires.

Figure 1.3(c) shows a simple cross-section of WREM media. In addition to the reflective layer there are three other layers deposited on the substrate. Full details of these layers will be given later, but for the present only the magneto-optic layer will be described. This magnetic layer can be magnetized by a small external vertical magnetic field when the focussed light spot heats it to a certain critical temperature. At this critical temperature the magnetic properties are so weak that it can be easily magnetized or remagnetized and the arrows on the diagram indicate the typical direction of magnetization of different parts of the film. The direction of magnetization can be detected by the *polar-Kerr effect* (see Chapter 2).

As with WORM media there are two laser powers: high power for writing or erasing; low power for reading.

## 1.3 The optical head

The basic components of the optical head are shown in Fig. 1.4.

- semiconductor laser
- beam-splitting prism
- focus lens
- photodetector

These same components are used in all optical recording systems. They were developed first for the CD audio system and are now manufactured in large volumes for low prices. This book does not deal with domestic audio systems but covers the commercial requirements of data retrieval and processing.

The semiconductor laser emits infra-red light at a wavelength of about 800 nm. This light passes through the beam splitter and then is focussed on to the media by the focus lens. The reflected light passes back through the focus lens and is deflected by the beam-splitting prism to the detector. The photodetector responds to the infra-red light and produces an electrical signal for data processing purposes.

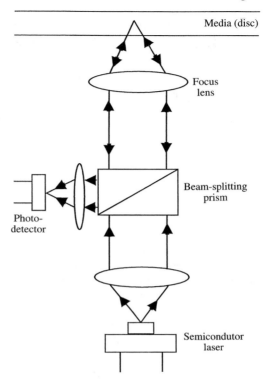

**Fig. 1.4** The basic components of the optical head.

# 1.4 Digital coding

Figure 1.5 illustrates one modulation code for a compact disc track. In this simple digital code the pit length represents the number of ones, and the length of the land between the pits is proportional to the number of zeros.

All three types of optical recording systems use digital coding and this makes them compatible with the majority of computer systems.

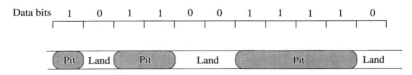

**Fig. 1.5** One modulation code for a compact disc track.

## 1.5 Comparison with conventional recording systems

To complete this introduction a simple comparison of the optical hard disc with the Winchester magnetic disc will be made. Table 1.1 summarizes this.

**Table 1.1**
**Comparison of optical discs with magnetic media**

|  | Magnetic | Optical |
| --- | --- | --- |
| Head positioning | Servo-positioning | Track-positioning |
| Data density | Medium | High ($\times$ 5) |
| Read/write head | Contacting | Non-contacting |
| Portability | Fixed | Removable |
| Error correction? | No | Yes |

First, the optical disc is generally the only disc in the drive, and hence a track is formed in or on the substrate before the recording layer is deposited. This track or groove enables the coarse positioning of the light spot. For the Winchester drive there is generally a spare surface, particularly in a multi-platter disc drive, which is used as the servo disc. Data is recorded on the servo disc so that the servo head can be used to position the other data heads.

Second, the density of data on the optical disc is about five times higher than on a magnetic Winchester disc. This higher density is owing to the limitation of the magnetic recording head in the Winchester drive. The head has to be a certain physical size in order to write data on the disc.

Third, a major advantage for optical recording is that there is no head contact. The head is about 1 mm away from the disc. For the Winchester drive the head lies on the disc whenever the computer is switched off. Taking off and landing the head is very hazardous with the Winchester, and because of the competition with optical recording the head flying heights are so low, that debris on the surface of the media can cause head 'crashes'. This means that the Winchester drive has to be assembled and sealed in clean-room conditions and the media can never be removed from the drive once it has been installed.

This leads on to the fourth point. The optical media can be removed from the drive and be protected in a cartridge or jewel box (CD-ROM). The portability of the media means that media can be sold or exchanged. This is particularly convenient when companies are working on a large number of industrial sites, and sharing information and software is essential.

Fifthly, error correction was included in optical drive systems from the beginning and this meant that the media could be of average quality, and therefore of very low price. For Winchester media the quality requirements are so strict that only about ten defects per side are allowed. This makes it comparatively difficult to produce high quality media at an economical price.

Finally, the CD audio revolution in 1990 and 1991 has increased the volume of CD drives manufactured to such an extent that as a result the price of optical recording drives is now competing successfully with the Winchester drive. Consequently, CD-ROM, WORM, and WREM optical systems all have an assured future for, at least the next ten to twenty years.

# Basic optics

2

## 2.1 Introduction to visible light and polarization

Light is a tranverse electromagnetic wave in which the electric and magnetic fields are perpendicular to each other and to the propagation vector $k$. Figure 2.1 shows a plane polarized light wave. The electric $E$ vector remains in a two-dimensional plane at right angles to the magnetic $H$ vector plane. The wavelength, $\lambda$, of the light is defined as:

$$\lambda = c/nv = \lambda_0/n$$
$$= 2\pi/k = 2\pi c/nw$$

(2.1)

where  $\lambda$ = wavelength (m)
$c$ = velocity of light (ms$^{-1}$)
$n$ = refractive index
$v$ = frequency (Hertz)
$\lambda_0$ = wavelength in vacuum (m)
$k$ = wave vector (radians m$^{-1}$)
$\omega$ = angular frequency (radians s$^{-1}$)

**Fig. 2.1** A plane polarized light wave showing electric, $E$, and magnetic, $H$, vectors.

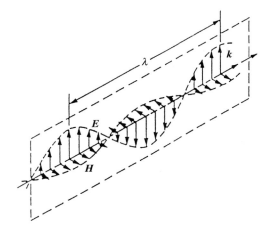

**Question**

> What is the velocity of light in glass with a refractive index of 1.5?

**Answer**

> Given that from the equations in (2.1) the velocity of light in a vacuum is calculated to be $3 \times 10^8$ ms$^{-1}$ you can work out that the velocity of light in glass is $2 \times 10^8$ ms$^{-1}$.
>
> Visible light forms only a very small part of the electromagnetic spectrum, as Fig. 2.2 shows. Current laser sources used in optical recording systems produce infra-red light with a wavelength of approximately 800 nm, just above the 1 μm point shown in Fig. 2.2.

# 2.2 Refraction and reflection of light

The velocity of light in a vacuum, $c$, is approximately $3 \times 10^8$ ms$^{-1}$. In glass or plastic the light slows down and the ratio of $c$ to the velocity in the glass or plastic is called the **refractive index**, $n$, of the glass or plastic.

*Snell's law* relates how the light direction changes when it slows down when passing from air to a more dense medium like glass. Figure 2.3 illustrates how the light beam is refracted as it slows down on entering glass. This refraction bends the light so that the angle of the refracted or transmitted ray to the normal $\theta_T$ is less than the incident light angle $\theta_I$.

These two angles are related by Snell's law:

(2.2)
$$n_1 \sin \theta_I = n_2 \sin \theta_T$$

where $n_1$ and $n_2$ are the refractive index of air and glass respectively.

The reflected ray is also shown in Fig. 2.3 where:

(2.3)
$$\theta_I = \theta_R$$

that is, the angle of incidence is equal to the angle of reflectance, $\theta_R$.

Finally, Snell's law does not apply for normal incidence, since the normal incidence ray does not bend. However, even at normal incidence there will always be some reflection of the light.

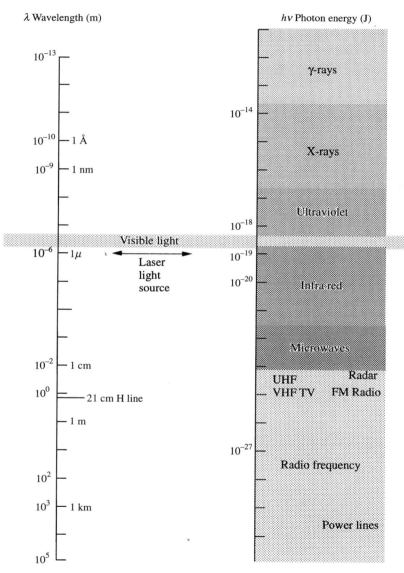

**Fig. 2.2** The electromagnetic spectrum showing relation of laser light sources to the visible and non-visible light sources.

**Fig. 2.3** The refracted light beam illustrating Snell's Law in which light changes direction as it passes through material of a different density.

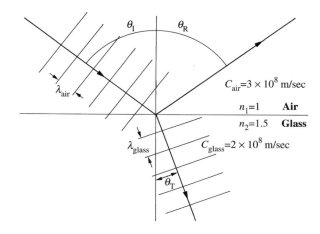

$C_{air}=3 \times 10^8$ m/sec

| | |
|---|---|
| $n_1=1$ | **Air** |
| $n_2=1.5$ | **Glass** |

$C_{glass}=2 \times 10^8$ m/sec

## 2.3 Lenses

For simplicity we shall discuss only one type of lens, the biconvex one. This lens is shown in Fig. 2.4 as part of an optical recording system.

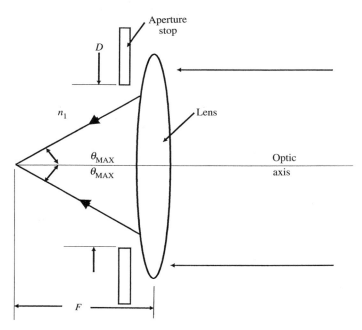

**Fig. 2.4** The biconvex lens focusing a parallel light-beam as used in optical recording systems.

The function of the lens is to focus the parallel beam to the smallest spot possible on to the media. The smallest spot occurs at the focal length distance $F$ from the lens. Since this focussing lens has to be held in position, the lens support acts as an aperture stop and limits the amount of the lens that can be used to focus the light. In other words $\theta_{MAX}$ defines the angle that the cone of light subtends with the optic axis for a circular aperture with stop of diameter $D$. The numerical aperture, $NA$, of the system is:

(2.4)
$$NA = n_1 \sin \theta_{MAX}$$

The $f$ number of the system is the ratio of the focal length, $F$, to the diameter, $D$, and is given by

(2.5)
$$f = \frac{F}{D} \approx 2NA \text{ for small } \theta_{MAX}$$

Two further definitions are needed for an optical recording system and these are the light spot width and the depth of focus. First, the minimum light spot width, $d$, is approximately:

(2.6)
$$d \approx 0.5\lambda / NA$$

## Question

What is the typical spot size achieved with optical recording?

## Answer

For a wavelength of 800 nm and $NA = 0.5$ then $d = 0.8$ $\mu$m.
The depth of focus, $z$, is given by:

(2.7)
$$z = \pm \lambda / 2(NA)^2$$

Using the same values for $\lambda$ and $NA$ as given above, then $z = \pm 1.6$ $\mu$m.

## 2.4 Diffraction

The optical disc grooves or tracks or pits act as a grating and split the incident light wave up into a number of diffraction orders. Figure 2.5 clearly shows the first orders on either side of the zero order which corresponds to the reflected light ray, with the reflectance angle equal to the incident angle (eq 2.3).

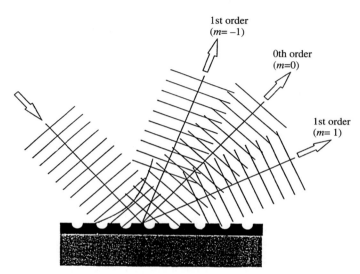

**Fig. 2.5** Diffraction from a one-dimensional grating representing first-order diffraction with optical disc tracks.

The diffracted angle beam rays are at an angle $\theta_m$, which is defined as:

$$\sin\theta_m = \sin\theta_I + m\lambda/p$$

(2.8)

where $m$ is the order number and $p$ is the period of the grating.

In practice, diffraction also occurs from the objective lens system shown in Fig. 2.4. This is because the lens acts as a circular aperture source, and this produces Fraunhofer **diffraction rings** around the central light spot. These rings are referred to as Airy rings, after Sir George Biddell Airy (1801–1892), the Astronomer Royal of England. Figure 2.6 shows the Airy pattern. The height of the first ring is only 1.8 % of the central spot intensity and the second is only 0.4 %.

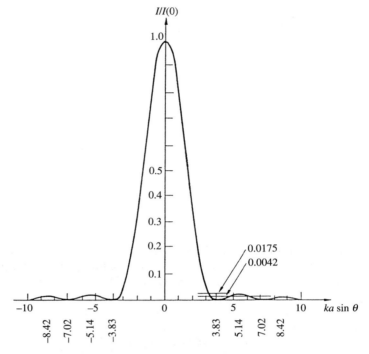

**Fig. 2.6** The Airy light spot intensity pattern which occurs with the objective lens in an optical recording system.

## 2.5 Polarization and birefringence

An optical component that polarizes light so that the $E$ vector stays in one plane, as in Fig. 2.1, is called a **linear polarizer**. There are four types of polarizer:

- selective absorption;
- birefringence or double refraction;
- reflection;
- scattering.

Figure 2.7 shows how the polarizer works. For simplicity, only two $E$ vector directions at right angles are shown in the light source. The polarizer absorbs one $E$ vector selectively and passes the other without absorption so that plane polarized light results.

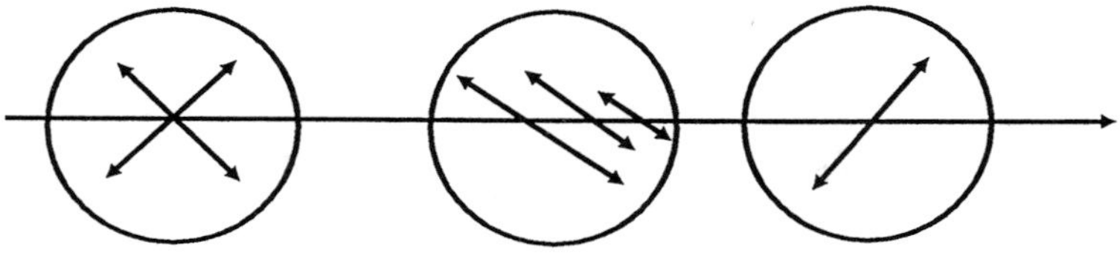


**Fig. 2.7** How a polarizer operates in selectively removing ***E*** vectors in all directions but one, to provide a polarized light source.

A material which displays two refractive indices is called birefringent.

Figure 2.8 shows a birefringent calcite crystal. Two light beams which are both plane polarized are produced. The ordinary ray, o-ray, passes through the calcite in the same direction as the incident ray; whereas the extraordinary ray (e) has a different path because the refractive index of the calcite varies with the polarization of the light beam. The optic axis is at right angles to the o-ray. Calcite crystals of this type can therefore be used in polarizing prisms.

Birefringence occurs in polycarbonate substrates which are most commonly used in optical recording. This will be dealt with in Chapter 5 when substrates will be described.

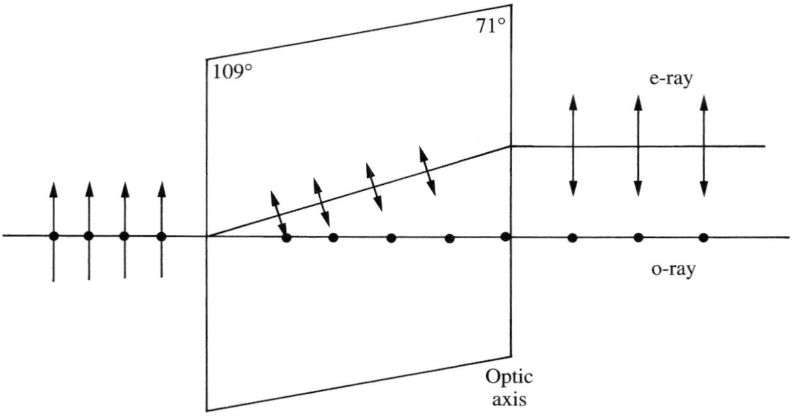


**Fig. 2.8** A birefringent calcite crystal which produces dual plane polarized light beams.

Wave plates or retardation plates can be used to change the polarization from plane to circular or vice versa, and can also rotate the polarization. Figure 2.9(a) shows how a half-wave plate rotates the plane polarized light through ninety degrees, and Fig. 2.9(b) illustrates how clockwise circularly polarized light (in which the $E$ vector rotates through 360° over one wavelength of the light) changes to anticlockwise polarized light after passing through the half-wave plate.

**Fig. 2.9** A half-wave plate with an incident light beam (a) plane polarized and (b) circularly polarized, showing changes in polarization.

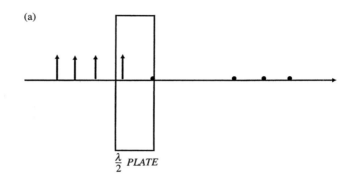

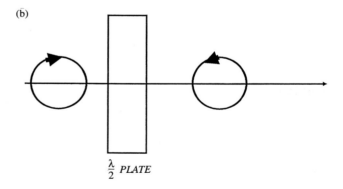

A quarter-wave plate, on the other hand, can change plane polarized light into circularly polarized light, as Fig. 2.10 illustrates. The orientation of the plane polarized light determines whether (a) clockwise or (b) anticlockwise circularly polarized light is produced.

**Fig. 2.10** A quarter-wave plate converts plane polarized light to circularly polarized light (a) clockwise and (b) anticlockwise rotation.

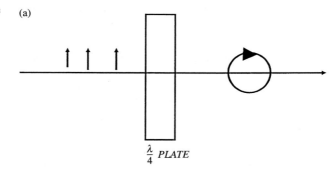

(a)

$\frac{\lambda}{4}$ *PLATE*

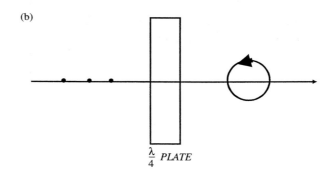

(b)

$\frac{\lambda}{4}$ *PLATE*

## Question

How can a quarter wave plate be used to prevent light being reflected back into the laser with the optical system shown in Fig. 1.2 using media with a final quarter wavelength optical coating and a reflective layer?

## Answer

Put the quarter wave plate between the laser and the first lens. This converts the plane polarised light to a circularly polarised beam which on reflecting from the media effectively passes twice through a quarter wave coating. This rotates the $E$ vector of the light wave through 90° so that it is reflected by the beam splitter on to the detector. Any light passing through the beam splitter will be out of phase by 90° with the light from the laser, and will have very little effect on the laser since all the waves from the laser are in phase.

## 2.6 The polar–Kerr and Faraday effects

Magneto-optic materials such as those used in WREM erasable media all show a *polar–Kerr* or *Faraday effect* when polarized light is reflected (polar–Kerr) or transmitted (Faraday) through them. In both cases the *E* vector is rotated upon reflection or transmission.

Figure 2.11 shows the Faraday effect with the light beam in the same direction as the magnetization *M* of the sample. The plane polarized light rotates as it passes through the sample and the rotation angle is called the **Faraday rotation**.

**Fig. 2.11** Rotation of plane polarized light when passing through magneto-optic material due to the Faraday effect.

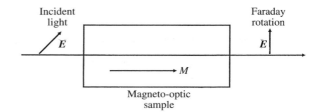

It is a similar case with the polar–Kerr effect, rotation of the *E* vector in plane or circularly polarized light on reflection from a magneto-optic film occurs. Figure 2.12 illustrates how the polar–Kerr rotation of the *E* vector depends on the direction of magnetization of a domain, and hence, as we shall see later in Chapter 9, a digital 'one' and 'zero' can be distinguished.

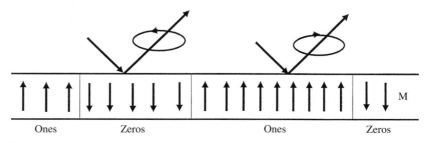

**Fig. 2.12** Circular polarization of plane polarized light after reflection from a magneto-optic layer due to the polar-Kerr effect which can be used to detect 'ones' and 'zeros' because of the change in rotation of the *E* vector.

# Lasers

**3**

## 3.1 Laser types

The gallium arsenide (GaAs) laser was invented by four groups of researchers at the same time. GaAs lasers are currently used in CD-ROM, WORM and WREM optical recording systems.

Lasing action at room temperature in a semiconductor material is relatively simple to achieve provided that:

- direct-gap semiconductors are used;
- non-radiative centres within the GaAs are kept to a minimum.

Direct-gap compounds and their alloys are essential because the recombination probability, $B$, to light emission centres is so much higher than in the indirect-gap materials as shown in Table 3.1.

**Table 3.1**
**Probability of recombination in semiconductor materials**

| Semiconductor | Gap Type | $B$ cm$^3$s$^{-1}$ |
|---|---|---|
| GaAs | Direct | $7.21 \times 10^{-10}$ |
| InP | Direct | $1.26 \times 10^{-9}$ |
| GaP | Indirect | $5.37 \times 10^{-14}$ |
| Si | Indirect | $1.79 \times 10^{-15}$ |

Non-radiative centres can be reduced to a considerable extent by using low temperature solution growth or metallo-organic chemical vapour deposition (MOCVD).

## 3.2 Criteria for lasing

The special semiconductor chip structure that is required to produce light-emitting 'lasing' action is shown in Fig. 3.1. Two sides are cleaved along the <110> crystal planes to give mirror finishes and two sides are cut either with a mechanical saw or an ultrasonic cutter. The laser is operated with a forward bias with the n-doped material connected to the negative side of the d.c. power supply.

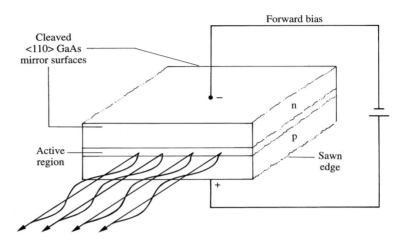

**Fig. 3.1** Schematic diagram of a gallium arsenide semiconductor laser junction showing the light-emitting surface along the <110> crystal planes.

When laser light emission occurs there is a significant increase in the quantum efficiency. This can be explained by using the electron energy band structure picture. For low applied voltage bias, as shown in Fig. 3.2(a), the excess electron density is low and spontaneous radiation occurs. Under these conditions we have the light emitting diode condition:

> The lifetime of spontaneous radiation is greater than the lifetime for non-radiative recombination

In this situation non-radiative recombination dominates and the internal efficiency is low. The internal efficiency, $\eta_i$, is given by:

$$\eta_i = \frac{1}{1 + \tau_{R_{Sp}} / \tau_{NR}}$$

(3.1)

**Fig. 3.2** Band diagram for (a) spontaneous emission under low bias conditions and (b) stimulated emission with high internal efficiency.

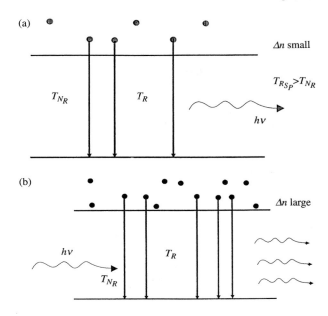

When the current is increased and the density of excess electrons becomes sufficient, the emission generated stimulates other electrons to decay with a much shorter lifetime so that:

> The lifetime for stimulated emissission is less than the lifetime for non-radiative recombination

The internal efficiency is now much greater. A plot of the radiation output power as a function of current density is shown in Fig. 3.3. The point at which a change in slope occurs is the threshold for lasing action. In practice, the **threshold current**, $J_t$, is not as clearly defined as this and it has to be determined by extrapolating the linear portion of the characteristic above the threshold print to intercept the current density axis.

The threshold for lasing action occurs when there is enough **stimulated emission** to produce a gain which is equal to the optical losses due to absorption and transmission at the Fabry–Perot mirror faces. Consider a **Fabry–Perot cavity** as shown in Fig. 3.4, with a length, $l$, and two partially-reflecting surfaces with a reflectance, $R$. The system is characterised by a gain per unit length, $g$, and a loss per unit length, $\alpha$. Consider a point in the cavity emitting a flux, $S_0$, in the forward direction.

**Fig. 3.3** Radiated power output for GaAs laser as the current density increases.

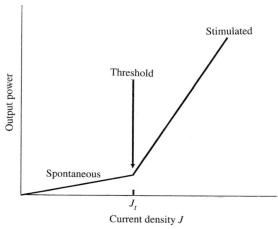

**Fig. 3.4** Falry-perot cavity demonstrating light-emitting conditions at the laser face.

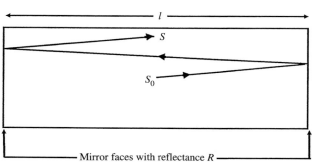

After travelling a distance *2l*, the flux *S* is given by:

(3.2)
$$S = S_0 R^2 \exp (2gl - 2\alpha l)$$

For stimulated emission to occur, the radiation at any point remains unchanged after the *2l* path, i.e. $S = S_0$. Thus the gain required for threshold of laser action $g_t$ is given by:

(3.3)
$$g_t = \alpha + \frac{1}{l} \ln \left( \frac{1}{R} \right)$$

The gain is proportional to the density of excess electrons and hence to the forward current. Thus the threshold current $J_t$ is proportional to *gt*, i.e. $gt = \beta J_t$. Substituting into eqn (3.3) this gives:

(3.4)
$$J_t = \frac{\alpha}{\beta} + \frac{\ln(1/R)}{\beta l}$$

## 3.3 Near- and far-field patterns

When the current is increased above the threshold value, first one and then more bright field spots appear at the line of intersection of the p-n junction within the Fabry–Perot mirror face, as shown in Fig. 3.5(a). By micrographing the front and back faces of a number of lasers it has been established that the spots appear in matched pairs at the ends of filaments running perpendicular to the Fabry–Perot mirror faces. A theoretical model of this filamentary formation suggests that the stability of these filaments is associated with the increase in refractive index which occurs when the local electron concentration is reduced by the shorter decay-time constant associated with stimulated emission. This is illustrated in Figure 3.5(b). The filament then acts as a **light pipe** and constricts the light along its length.

**Fig. 3.5** Near-field pattern of diffused large area laser junctions showing localized filament formation in regions of reduced electron density. (a) Filament formation and (b) excess electron density through a filament.

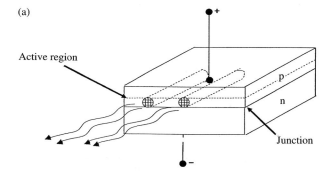

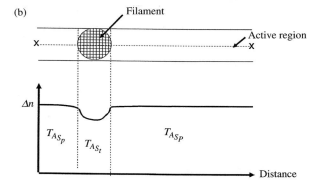

Filamentary lasing has been avoided by the use of the stripe geometry illustrated in Fig. 3.6. For a normal diffused junction the stripe width, $d$, is usually chosen to be the same as the filament width of approximately ten microns. For such a structure standing waves occur when the cavity contains an integral number of half wavelengths. For a cavity of length $l$ this condition is expressed by:

(3.5)
$$\frac{m\lambda}{2n} = l$$

where $m$ is an integer, $\lambda$, is the wavelength of the radiation in the semiconductor and $n$ is the refractive index.

**Fig. 3.6** Mesa stripe structure of a gallium arsenide laser diode to reduce filament formation.

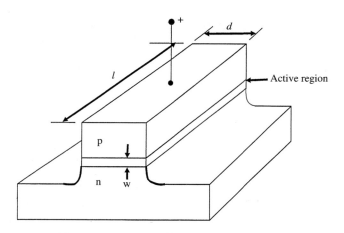

Typical values l=200 $\mu$m; d=10 $\mu$m; w=1 $\mu$m

The spacing between the modes is given by:

(3.6)
$$\Delta\lambda = \frac{\lambda^2}{2l\left(n - \lambda\dfrac{dn}{d\lambda}\right)}$$

In addition to longitudinal modes it is possible to propagate modes having a transverse component. If the sides of the junction are plane and parallel, a distant, $d$, apart, then the radiation emerging from the ends will show intensity peaks at an angle $\phi_N$, as shown in Fig. 3.7, where:

(3.7)
$$\sin \phi_N = n \sin\left[ \tan^{-1} \frac{d}{Nl} \right]$$

where $n$ is the refractive index, $N$ is the number of the mode and $l$ is the lower cavity length.

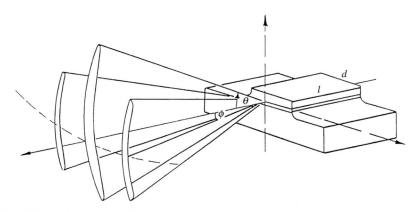

**Fig. 3.7** Far field pattern of a stripe laser junction exhibiting transverse propagation.

In the plane of the active layer the width of the beam at half maximum intensity $\Delta\phi$ is given by $\Delta\phi = \lambda/d$, where $d$ is the width of the active layer. In the plane transverse to the active layer, the radiation is also subject to single-slit diffraction. Thus the beams diverge with a much larger angle of half intensity $\delta\theta$, given by $\delta\theta = \lambda/w$, where the thickness of the active recombination layer is typically one micron. As expected, the beam intensity is highest when $\theta$ or $\phi = 0$. At half the maximum intensity $\theta = 20°$ and $\phi = 5°$ are typical values for standard gallium arsenide homojunctions. (A homojunction is one in which both the p, and the n side of the laser are made from gallium arsenide.)

## 3.4 The heterostructure laser

The infra-red laser used in optical recording systems is of the GaAs/AlGaAs double heterostructure type. Figure 3.8 shows the energy band edges under forward bias lasing emission conditions, with refractive index changes, and optical field distribution in homostructure, single-heterostructure and double-heterostructure lasers.

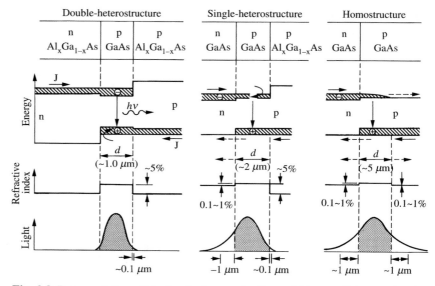

**Fig. 3.8** Representation of the band edges under forward bias conditions, refractive index changes, and electrical field distribution in homostructive, single heterostructive and double heterostructive lasers.

In the single-heterostructure, the penetration of the radiation into the p-type GaAlAs layer shown in Fig. 3.8 is minimal, owing to the refractive index change, and with the double-heterostructure penetration of the radiation, is further confined to the p-type GaAs active layer, thus increasing laser efficiency. The interface between the active p-type GaAs region, with the n and p-type ternary compounds, acts as a potential barrier to electron and hole penetration. This also improves the laser gain by increasing the density of excess carriers in the active region.

A typical laser characteristic for a commercial double-heterostructure GaAs/AlGaAs laser is shown in Fig. 3.9. The threshold current is only 14 mA. Figure 3.10 shows the far field angular intensity distribution when the laser power $P_o$ is 1.5 mW and confirms that it is an excellent light source.

## 3.5 CW operation

CW stands for continuous wave, and this is the way that the laser is operated in CD-ROM and WORM optical recording systems. Since the laser diode will slowly deteriorate with continuous operation the light emission intensity will also decrease. To avoid this, the light emitted from the back face is used to power a photodiode (PD) detector which automatically increases the laser

**Fig. 3.9** A typical laser characteristic for a double-heterojunction laser.

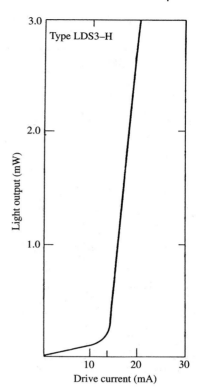

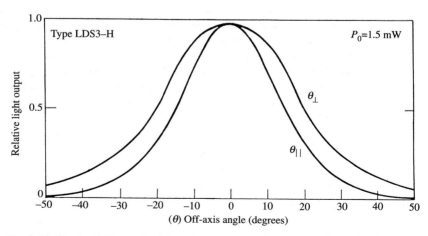

**Fig. 3.10** The far field angular intensity distribution for the laser shown in Fig. 3.9.

current when the output light power decreases and returns the power to a set value.

An example of a driving circuit for laser diode (LD) is shown in Fig. 3.11. $R_L$ stands for the load resistor.

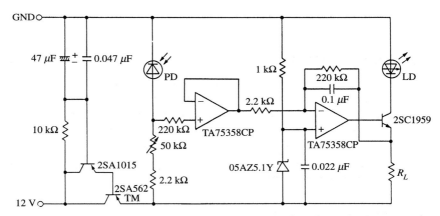

**Fig. 3.11** The driving circuit for a laser diode (LD) incorporating compensating photodiode (PD) to automatically increase laser current when light output drops.

### Question

What is the difference between a heterojunction and a homojunction gallium arsenide semiconductor laser? Why use a double heterojunction laser?

### Answer

A homojunction is just a gallium arsenide p–n junction with a refractive index variation of only 0.1 to 1 % at the depletion region boundary. (The depletion region is the region in which the electron or hole is subject to a potential gradient in the vicinity of the p–n junction. This voltage gradient depletes the region of electrons and holes.) A single heterojunction has a p-type gallium–aluminium arsenide region next to the p-type gallium arsenide region. (Hence there are three regions of doped semiconductor material.) The refractive index change at the gallium–aluminium arsenide junction is now 5 %. This change produces a wave-guiding effect and narrows the laser emission on one side.

For a double heterojunction there are two aluminium–gallium arsenide layers on either side of the p-type gallium arsenide. One is n-type and the other is p-type. The laser is therefore effectively narrowed on both sides and the beam is emitted from a one-micron region compared with 5 $\mu$ for the homostructure. Thus a very effective light guide (just like an optical fibre) is formed. (Figure 3.9 compares the three types of laser.)

# Photodetectors

4

## 4.1 Silicon detectors

Detectors that are used in optical recording systems are all made from silicon. Silicon detectors have a wavelength response both in the visible part of the spectrum and in the infra-red. The light from a gallium arsenide laser can be detected with an effective quantum efficiency of greater than ten per cent. The effective quantum efficiency is defined as the ratio of electronic current generated to the number of photons incident on the p–n or p–i–n extended junction (see below).

## 4.2 The p–n junction

The ideal abrupt p–n junction is shown in Fig. 4.1. In (a) the impurity profile shows an ideal uniform acceptor concentration, $N_A$, on the p side of the junction and an ideal uniform donor concentration, $N_D$, on the n side of the junction. Figure 4.1(b) shows the electron energy diagram for an unbiased diode. The Fermi level is constant throughout, and the potential barrier owing to the junction is $eV_d$. In Fig. 4.1(c) under reverse bias the barrier has increased to $e(V_a + V_d)$, where $eV_a$ is equal to the energy difference in the **Fermi levels** on both sides of the junction. Provided the junction width and thickness of the device in the p or n region that is illuminated are correct, light can penetrate all through the junction.

There are three absorption processes which occur when the diode is under **reverse bias**:

- in the p region;
- within the junction;
- in the n region.

In each case, electron–hole pairs are generated, as shown in Fig. 4.2, and they all contribute to the photo current flow in the device.

The current responsivity as a function of wavelength is shown in Fig. 4.3 Wilson and Hawkes (1983). Responsivity is the current in amperes generated per watt of light power. Note the peak or strongest lasing emission line is at approximately 850 nm (8500 A°), indicating what an excellent response the silicon detector has at the gallium arsenide laser emission wavelength. Also

**Fig. 4.1** The ideal abrupt p–n
junction: (a) uniform doping;
(b) electron energy diagram for an
unbiased diode; (c) electron energy
diagram under reverse bias.

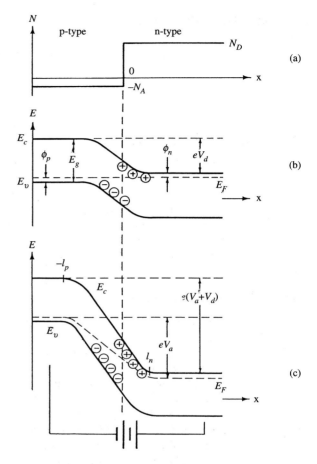

shown in Fig. 4.3 is the ideal photodiode responsivity with unit quantum
efficiency.

## 4.3 The p–i–n photodiode

The p–i–n photodiode is very popular in optical recording systems, since the i
intrinsic layer extends the **depletion layer** so that the frequency response and
quantum efficiency can be optimised. (It is called the depletion layer because
it is a region of very few free carriers, and electron and hole concentrations
are nearly equal.) Figure 4.4(a) shows a cross-section of the device Sze
(1985). An anti-reflective coating helps the light to couple into the p-layer,
and a silicon dioxide passivation layer insulates the metal contact to the
player from the intrinsic region i. The diode is connected in reverse bias
with a load resistor, $R_L$.

**Fig. 4.2** Electron–hole pair generation in a reverse biased photodetector diode.

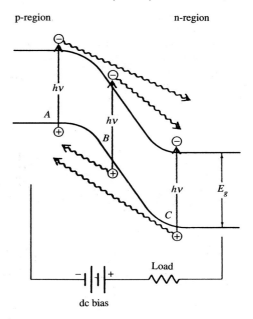

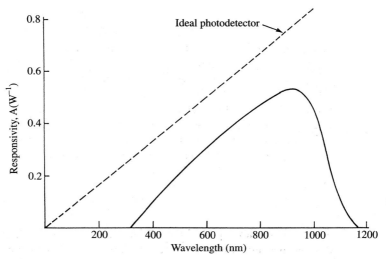

**Fig. 4.3** Responsivity variation with wavelength showing optimum and ideal power conditions.

**Fig. 4.4** The p–i–n
photodetector: (a) cross-section
(b) energy band diagram under
reverse bias; (c) carrier
absorption characteristic.

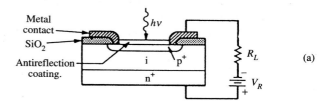

(a)

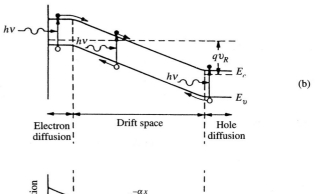

(b)

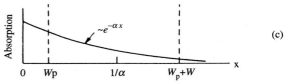

(c)

In Fig. 4.4(b) the electronic energy band diagram of the p–i–n diode under
reverse bias clearly shows this extension of the depletion region or drift
space. It is called drift space because electrons or holes generated in this
region will drift in the direction of the n-region for electrons, and towards the
p-region for holes. Once again, as in Fig. 4.2, the three types of photocurrent
or electron–hole pair generation are shown.

Figure 4.4(c), shows the light absorption extending throughout the i-layer
and dying out in the n-layer. The light decays exponentially:

(4.1)

$$I = I_0 e^{-\alpha x}$$

where $I$ is the absorption intensity variation as a function of $x$, the distance
through the diode, and $I_0$ is the intensity at $x = 0$, the light intensity at the
diode surface. $\alpha$ is the absorption coefficient of the silicon crystal.

Quadrant p–i–n detectors are used for fine focus control, as will be seen in Chapter 7.

A typical specification is as follows:

| | |
|---|---|
| Peak responsivity | 0.5 A/W |
| Dark current | 0.1 nA per quadrant ($V_R = 1$ V) |
| Capacitance | 45 pF ($V_R = 0$ V) |
| Response time | 15 ns ($V_R = 10$ V) |
| Reverse breakdown voltage | 60 V |
| Peak spectral response | 800 nm |
| Active area | 0.66 mm$^2$ per quadrant |

The response time is the time taken for the current output to reach a maximum with a constant energy light source falling on the detector. ($V_R$ is the voltage rating.) The dark current is the residual noise current on the device when the light is no longer shining on it and it is in complete darkness.

**Question**

How is the p–i–n photodiode used to detect laser light in an optical recording system?

**Answer**

The diode is reverse biased to increase the potential barrier. This increases the absorption of light and hence the responsivity of the photodiode. As Fig. 4.2 shows, electron-hole pairs (or photo-excited carriers) are generated within the depletion region as well on the p- and n-side of the junction. The i-layer has approximately equal numbers of electrons and holes, so this gives an extended depletion layer and hence increases detectivity further. However, if the i-layer is too thick the light will all be absorbed before it reaches the n-side of the junction, as shown in Fig. 4.4.

The antiflecting coating is a quarter of a wavelength thick, so that it acts as a polariser and converts all the plane polarised light into circularly polarised light, which then passes into the p-layer. The total thickness of the p-layer must be below one micron, since if it is over one micron the circularly polarised light will all be absorbed in the heavily doped p region (called p$^+$).

# 4.4 The position-sensitive detector (PSD)

The PSD can also be used in an optical recording system. There are two basic types: one-and two-dimensional PSDs. First, the one-dimensional type can be used to explain the operating principle.

Figure 4.5 shows the side view cross-section. Light generates carriers in the surface of the p-layer which are detected by two output contacts to give currents $I_1$ and $I_2$ respectively. If the light spot is a distance $x$ from the centre of the device then:

(4.2)
$$\frac{x}{L} = \frac{I_2 - I_1}{I_1 + I_2}$$

where $L$ is half the effective width of the detector surface, as clearly shown in the diagram.

**Fig. 4.5** Cross-section of a one-dimensional position–sensitive detector.

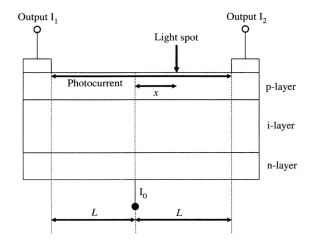

Secondly, the area detector has two pairs of top electrodes. If the photocurrents in the x-direction are $I_1$ and $I_2$ and the photocurrents in the y-direction are $I_3$ and $I_4$, then, as well as eqn (4.2) to define $x$, $y$ is given by:

(4.3)
$$\frac{y}{L} = \frac{I_4 - I_3}{I_3 + I_4}$$

This assumes a square device with identical spacings $2L$ between the $x$ and $y$ top surface electrodes.

# References

Sze, S. M. (1985) *Semiconductor Devices*, John Wiley, page 283.
Wilson, J. and Hawkes, J. F. B. (1983) *Optoelectronics: An Introduction*, Prentice-Hall, page 313.

# Media substrates

## 5.1 Disc substrate materials

The following substrate materials have been considered for optical discs:

- Polycarbonate
- Glass
- PMMA

- Polyethersulphone
- Polyimide
- Amorphous polyolyphine
- Epoxy
- PC/PS

Of these eight alternatives the first three have been highlighted since they are the only ones that are in common use in optical data recording.

Polycarbonate is an attractive plastic substrate for optical recording because it has a high optical transparency, can be moulded to a very high precision and is relatively cheap. The excellent moulding characteristics have led to it being adopted for all CD-ROM discs and as the main competitor to glass for 130 mm WORM and WREM discs.

Glass, although it is currently more expensive than polycarbonate, has excellent optical and mechanical properties. Its mechanical strength makes it particularly suitable for 300 mm WORM discs.

PMMA, (polymethyl-methacrylate) with a low birefringence, is particularly suitable for the large video discs, but it suffers from moisture absorption and low heat resistance. Moreover, as multi-media CD-ROM versions become more popular, the video-only disc will gradually be phased out and so will PMMA.

Of all the possible other alternatives, amorphous polyolyphine looks very attractive, and it will be featured with polycarbonate and glass when specific properties are discussed below.

Next we need to consider what general properties are required by substrate materials.

## 5.2 Optical properties

Since light always passes through the substrate, the key properties are as follows:

- high transmission;
- good surface optical quality;
- low birefringence.

The light beam actually passes through the substrate twice since it is reflected either from the aluminium mirror layer (for CD-ROM and WREM) or from the data recording layer (for WORM).

Transmissions of greater than 90 % can be achieved for both glass and polycarbonate provided that the bubbles, inclusions, surface defects and deep scratches are all kept to a minimum. Obviously this transmission will have an effect on the reflectivity that is achieved, and hence on the signal-to-noise of the recording system.

Good surface quality is essential for high transmission and reflectivity of the mirror or data layer that is deposited on the substrate. In the case of polycarbonate, this can be achieved with the correct moulding conditions (Chapter 7). For glass, polishing is required to give an optical quality surface flatness.

Birefringence depends critically on the thickness of the substrate. The lateral or longitudinal birefringence for polycarbonate increases as the square of the angle of the incident light from the normal increases.

From Fig. 5.1 it is obvious that for polycarbonate substrates the focusing of the light is a problem, and this produces distortion of writing, reading and erasing with the laser beam. Increased optical feedback into the laser can also occur. However, as is shown below, polycarbonate has now been improved to such an extent that the birefringence at normal incidence, $BR_z$, is quite low.

## 5.3 Mechanical properties

Three key properties are required:

- thickness uniformity;
- surface evenness;
- strength.

**Fig. 5.1** Birefringence in a polycarbonate disc substrate at near-normal incidence of laser beam.

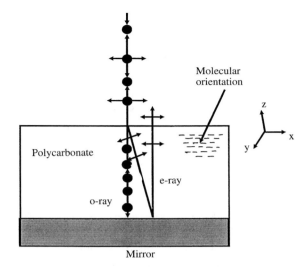

Thickness uniformity, as well as affecting the birefringence, causes optical interference fringing if there is a wedge structure, and imbalance which will limit the speed at which the disc can be rotated.

Surface evenness is affected by warping, waviness and roughness. Roughness describes the local thickness variation, whereas warping and waviness concern the whole disc. The roughness can be determined with a lightly loaded stylus on a surfometer. Waviness and warping are also called axial runout. Total indicated runout, TIR, can be measured at a fixed radius as the disc rotates. The TIR will change as the disc speed increases, especially for polycarbonate substrate since the disc flattens out owing to the centrifugal forces.

Strength is obviously important and stress cracking during manufacture must be avoided. Fracture can be a problem with a single glass substrate that is not hardened, but in WORM designs two hardened glass discs are bonded together, so 'double glazing' which is very strong is produced.

# 5.4 Thermal requirements

This will depend on the application. For CD-ROM discs the temperature properties are related to the manufacturing process and these will be dealt with in detail in Chapter 7. In the case of WORM and erasable media, short temperature excursions are experienced as the laser heats a single spot. With WORM discs, where high temperatures well above 200°C are required to melt the film, the laser pulse length becomes quite critical.

Rheological (or heat flow) properties are also important, since they determine the profile of the heated spot as a function of time.

## 5.5 Stability requirements

There are three things that will help the substrate to support long life data integrity within the recording media:

- low moisture content and transmission;
- chemical inertness;
- imperviousness to oxygen.

Moisture was initially a serious problem with polycarbonate. However, higher purity variations have now been produced in which the water content is tolerable. PMMA always suffers from moisture absorption.

Glass is the most reliable material from the stability point of view, since it has a very low moisture content and is impervious to oxygen. It does contain sodium and potassium ions, so either a low ion content glass is made or a barrier layer is deposited first before the optical recording layers are coated.

## 5.6 Other properties

There are several other key issues for substrates that should be mentioned:

- cost;
- adhesion properties;
- ease of manufacture;
- mechanical tolerances.

Cost is vital, particularly as the optical media products are beginning to establish themselves in the general market. Polycarbonate is cheaper than glass at the moment and easier to manufacture, but this could all change when float glass, which is used for window manufacture, takes over from plate cast glass.

Adhesion of films to polycarbonate are better at higher temperatures. This is also the case with glass, but sometimes a bonding layer like chromium is used to improve the adhesion to glass.

Mechanical tolerances are very important and these can easily be achieved in the mould design with polycarbonate. Glass requires cutting and polishing and this is time-consuming and expensive.

## 5.7 Polycarbonate (PC) substrates

At this point the properties of polycarbonate will be summarised. First, the chemical formula of polycarbonate is shown in Fig. 5.2. This long chain polymer orientates with the polycarbonate backbone along the x-direction, as shown in Fig. 5.1. This orientation causes the birefringence.

**Fig. 5.2** The chemical formulae of polycarbonate commonly used in disc manufacture.

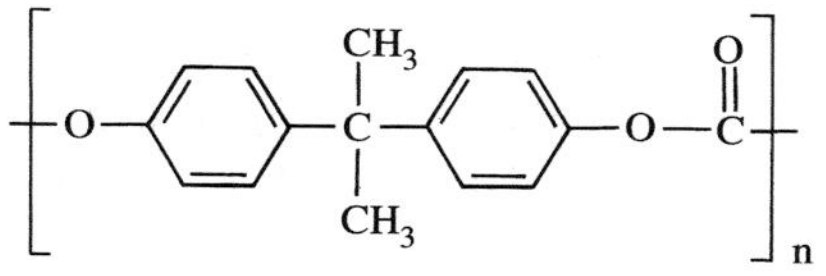

Copolymer formation, by adding an additional monomer to the polycarbonate, has been suggested as a way of reducing the birefringence. Further work is required to optimise these modifications of polycarbonate.

Another approach has been to modify the injection moulding technique, so that a combination of injection and compression is used to randomize the molecular orientation of the polycarbonate which also reduces the stresses within the disc. Once again, this is still under development, but it is expected to improve the material further.

Now here are a few more facts about polycarbonate:

| | |
|---|---|
| Density | $1.2 \text{ g cm}^{-3}$ |
| Glass Transition Point | $128°C$ |
| Tensile strength at break | $560 \text{ kg cm}^{-2}$ |
| Rockwell Hardness | 45 |
| Pencil Scratch Test | HB |
| Optical Transparency | 93 % |
| Refractive Index | 1.59 |
| Vertical Birefringence | < 60 nm |
| Water Absorption | 0.2 % |

## 5.8 Amorphous polyolyphine

Since polyolyphine promises to be a possible replacement for polycarbonate a comparison of the properties shown above follows:

| | |
|---|---|
| Density | 1.05 g cm$^{-3}$ |
| Glass Transition Point | 150°C |
| Tensile strength at break | 590 kg cm$^{-2}$ |
| Rockwell Hardness | 75 |
| Pencil Scratch Test | 2H |
| Optical Transparency | 93 % |
| Refractive Index | 1.55 |
| Vertical Birefringence | < 20 nm |
| Water Absorption | 0.01 % |

From the point of view of hardness, water absorption, birefringence and glass transition temperature, amorphous polyolyphine looks excellent. All that remains to be achieved is the difficult task of optimising the injection moulding process to produce a high quality CD-ROM finish or grooved WORM or grooved WREM substrate with this novel plastic.

## 5.9 Glass

Glass is a standard substrate for WORM but is not normally used for CD-ROM. For certain WREM applications glass is also being accepted.

The general properties of drawn sheet glass substrates are as follows:

| | |
|---|---|
| Density | 2.47 g cm$^{-3}$ |
| Youngs Modulus | 715 GNm$^{-2}$ |
| Hardness Vickers | 4.8 GNm$^{-2}$ |
| Optical Transparency | 88.5 % for 1.2 mm |
| Refractive Index | 1.504 |
| Vertical Birefringence | < 40 nm cm$^{-1}$ |

A full specification for glass, including thermal and chemical properties, can be obtained from Pilkington plc.

Acceptable roughness of 5 nm can be achieved with unpolished glass. The surface roughness, peak-to-valley, is measured using a Talystep with a 0.1 μm radius chisel stylus over a sample length of 200 μm. A polished chemically-strengthened glass substrate has a surface roughness of less than 3 nm and the parallelism and flatness is often a factor of ten better than for unpolished substrates.

Float glass, which is cheaper since it is produced with a cheap process for windows in buildings, has very similar optical and mechanical properties. The refractive index is slightly higher at 1.517, but the transmission is identical, since it is 88.5 % at 800 nm wavelength. The birefringence is also less than 8 nm for an optical substrate.

## 5.10  PMMA and comparison with PC and glass

Table 5.1 compares the properties of the three substrate materials currently used in optical recording.

**Table 5.1**
**Comparison of common disc materials**

| Property | Polycarbonate | PMMA | Glass |
|---|---|---|---|
| Birefringence | Average | Good | Good |
| Manufacturability | Good | Good | Fair |
| Water absorption | Average | Poor | Good |
| Density | Good | Good | Fair |
| Operation temperature | Good | Average | Good |
| Abrasion resistance | Fair | Good | Good |
| Impact strength | Good | Average | Fair |
| Cost | Good | Good | Fair |

This table shows that at the present time glass is unfavourable when all these properties are taken into account. PMMA (Polymethyl methacrylate) is very similar to PC but the water absorption is so bad that long-term corrosion is a serious problem, unless the PMMA has an additional water barrier layer. This will put up the cost, which is currently lower than polycarbonate.

So polycarbonate must be the favourite material and undoubtedly will continue to be improved to such an extent that CD-ROM and erasable WREM media manufacturers will use it in preference to glass. For WORM, on the other hand, because the main application is archival storage over many years, glass will be favoured.

Glass will become cheaper as float glass takes over from plate or cast glass methods, so it could well move into more WREM markets. However, until etching of grooves in glass is perfected, the cost will still be too great for general applications because of the additional cost of putting down a polymer structure to define tracks on the glass.

## Question

Why is polycarbonate the preferred substrate for optical recording at the present time (1993)?

## Answer

Reference to Table 5.1 shows that polycarbonate (PC) is easy to manufacture and the cost is cheaper than glass. PMMA, on the other hand, absorbs water to such an extent that the films on it are easily corroded. The abrasion resistance (AR) is poor for polycarbonate but in most cases, except for CD-ROM, the disc is always in a caddy and the optical head objective lens is 1 to 3 mm above it, and is never touched. Even for the CD-ROM most drives now require a caddy and some manufacturers supply the disc already in a caddy. Slight scratches are no problem, since the error correction that is built into every disc sector corrects errors of this nature.

# Mastering | 6

## 6.1 Introduction

Creation of high-quality prerecorded masters is required for the production of all 120 mm CD audio, video and data discs. In this chapter the emphasis will be on mastering CD discs, but much of what is said will also apply to the mastering of the 300 mm video discs and to track masters for WORM and WREM polycarbonate substrates.

## 6.2 Process review

To give an overview at the beginning the full process is shown in Fig. 6.1. A polished glass substrate is cleaned and dried after inspection and then an adhesive coat is used to promote good adhesion of the spun-on photoresist coat which is sensitive to ultraviolet light. After inspection of the photoresist, it is baked to harden it. A blue laser containing UV components is then used to expose a pit pattern on the photoresist. The exposed areas are washed out by developing the resist as in PCB production and the remaining photoresist is metalised with silver in a sputter coating machine. Using a special disc player for glass masters to 'playback' the data (audio or video or a combination of these) the quality of the recording can then be evaluated.

## 6.3 The glass master substrate

The glass substrates for CD mastering are 200 mm in diameter and normally polished on both sides. The float glass that is used is ground and polished using proprietary techniques. After polishing, a metal holding button is glued to one side in the centre of the substrate. The outline specification at this stage will be:

| | |
|---|---|
| Outside diameter | $200.00 \pm 0.05$ mm |
| Parallelism | 0.005 mm |
| Flatness | 0.01 mm |
| Thickness | $5.00 \pm 0.025$ mm |

After preparation the glass master is sealed in nitrogen in special packaging in a clean air Class 100 cabinet.

**Fig. 6.1** Steps in the mastering process for a CD disc using glass as the master substrate.

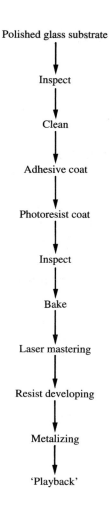

Polished glass substrate

Inspect

Clean

Adhesive coat

Photoresist coat

Inspect

Bake

Laser mastering

Resist developing

Metalizing

'Playback'

## 6.4 Resist coating

The glass substrate is unpackaged in a clean air cabinet which is linked to a second clean air cabinet in which the adhesive and the photoresist are spin-coated on to it. It then goes into a third clean air cabinet, where the resist is baked at 100°C in a convection oven for 20 to 30 minutes. The resist thickness after baking is uniform and in the range 0.08 to 0.12 μm.

## 6.5 The laser beam recorder

The third clean air cabinet, where the resist is baked, is linked to the clean room in which the laser beam recorder is situated.

Figure 6.2, Putt (1990), shows the argon ion laser beam recorder without the focusing system, which will be discussed separately below. The 458 nm or 488 nm line of the argon ion is used, and this allows a focussed spot of 0.5 μm diameter to be achieved. The laser power can be in the range 25 to 50 mW. The exposure of the photoresist is controlled by chopping the laser beam with an electro-optic or acousto-optic modulator. The modulator is controlled by an encoder driven from the master data tape (see Chapter 7). These modulators work in the low MHz range (typically at 25 MHz). The EOE, the electro-optic-electric detector which checks that the modulator is operating correctly, must be a high frequency type. The pit volume depends both on the laser light intensity and the tangential velocity of the disc at the light spot. For constant linear velocity, as in the CD-ROM case, the average light intensity must be kept constant across the disc as the disc slows down or speeds up. (The CD-ROM disc runs at 200 rpm when the laser is focused in the centre of the disc and 530 rpm at the outer diameter.) In order to keep the intensity constant, as Fig. 6.2 (Putt, 1990) indicates, an intensity detector is used with a feedback circuit to change the intensity control, which can be an acousto-optic modulator. Average exposure times for each pit are 100 ns.

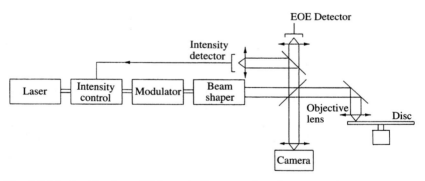

**Fig. 6.2** The 'writing beam' light path of an argon laser beam recorder.

The beam shaper shown in Fig. 6.2 is a beam expander that magnifies the Gaussian cross section of the laser beam in order to fill the objective lens aperture more homogeneously. Also shown in the figure is a CCD camera optical system, which is used to observe the quality of the written spot.

Focussing during mastering is very critical and so a separate laser system, as shown in Fig. 6.3, is used to control it. A 5 mW helium–neon laser is used, and a focus error detector controls a voice coil actuator or air-bearing motor, which holds the objective lens. Usually a quandrant detector is used, in the same way as with the compact disc focus control system described in Chapter 7.

**Fig. 6.3** The focusing beam light path on a laser beam recorder.

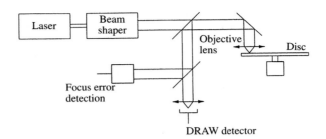

Mastering is carried out in class 100 clean room conditions. The recording objective lens has a numerical aperture of 0.75 and is supported on an air-bearing lens motor. During recording (or exposure of the pits) the objective lens is moved sideways by a linear motor driven by a specially-designed precision leadscrew assembly.

The glass master disc is supported by an air-bearing spindle driven by a phase-locked servo motor. The complete transfer assembly is carried on a specially-designed high-rigidity casting, which contains an integral air suspension system for vibration isolation. The whole transport is also completely enclosed in a housing maintained at a positive pressure with Class 100 filtered air.

The electronic master-control system is microprocessor-based so that the entire recording process is automatic. Recording a disc with 600 MBytes of data takes approximately one hour.

The final pit shape is determined by the light spot shape. This is because there is no post-bake of the exposed resist to enhance diffusion. Nor is there evidence of photon-stimulated diffusion.

The pit depth is nearly always made equal to the resist thickness, so that the developing washes out the pits cleanly to the glass surface, as shown in Fig. 6.4. The pit spacing, $p$, is related to the rotational frequency of disc, $f_R$, and the track radius, $r$:

(6.1)
$$p = \frac{2\pi r f_R}{f_t}$$

where $f_t$ is the temporal frequency of the laser pulse. With $f_t = 7.9$ MHz, $r = 55$ mm and $f_R = 25$ Hz then $p = 1.1$ $\mu$m. For a CD master, the track separation on the master is controlled to $1.6 \pm 0.1$ $\mu$m.

**Fig. 6.4** The pit width, $d$ and the pit spacing, $p$, on a glass master disc after developing the photoresist.

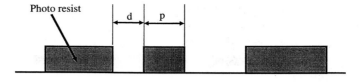

To calculate the pit, track or groove profile the Airy light spot calculation can be used:

$$I(r) = I_0 \left( \frac{2J_1(2\pi NAr\lambda^{-1})}{2\pi NAr\lambda^{-1}} \right)^2$$

(6.2)

where $I_0$ is the intensity at $r = 0$ and $J_1$ is the Bessel function of the first kind and first order. The width of the central light spot is proportional to the wavelength and the numerical aperture ($NA$). As eqn (2.6) gave:

$$d \approx 0.5 \, \lambda NA$$

This is the diameter at half the maximum intensity of the central light spot. In practice, the diameter of the first dark ring, $d_{DR}$, is often more useful:

$$d_{DR} = 1.22\lambda(NA)$$

(6.3)

## 6.5 Testing

Finally, after silvering the master, there are two options. Either further optical phase measurements on the first diffracted orders can be carried out, or a master inspection player which measures the following can be used to test the disk:

- carrier to noise ratio;
- signal to noise ratio;

- carrier amplitude;
- peak to peak amplitude;
- signal drop outs.

## Question

Why are the pits produced in photoresist on a CD glass master one quarter of a wavelength deep?

## Answer

They have to be exactly one quarter wavelength deep (about 20 nm for a gallium-arsenide laser) so that reflected light from the bottom of the pit is cancelled and appears as a dark spot of the 'land' area between the pits. A half waveplate is placed between the CD master and the beam splitter so that the light passing through it twice and reflecting from the bottom of the pit is exactly 180° out of phase with the original wave and so cancels it out. In the land case, assuming near-perfect reflection, the phase shift is only 90°, so cancellation does not occur.

# References

Bouwhuis, G., *et al.* (1985) *Principles of Optical Disc Systems*, Adam Hilger, page 196–201.
Putt, P. L. M. (1990) *MRS Bulletin*, April, page 55.

# Making the CD-ROM book

7

## 7.1 Principles of reading with lasers

CD-ROM is a read-only memory which is replacing the printed word for a wide variety of applications. Consequently, this chapter is all about making the CD-ROM book.

Since the average book page contains 3 Kbytes of data per page, the CD-ROM with 600 Mbytes is equivalent to 200,000 book pages.

Floppy discs, microfilm and microfiche all store very small amounts of data in comparison with the CD-ROM discs.

Why can the CD-ROM store so much data? The answer lies in the pit sizes and the track spacings that are used to store digital data. Figure 7.1 shows the

**Fig. 7.1** The CD-ROM pit sizes and track spacing which hold 600 Mbytes of data per disc.

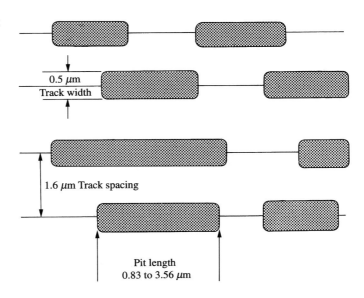

0.5 μm
Track width

1.6 μm Track spacing

Pit length
0.83 to 3.56 μm

CD-ROM pit sizes looking down from the top of the disc. The pit is approximately 0.5 $\mu$m in width, and from 0.8 to 3.6 $\mu$m in length, depending on the number of ones or zeros that are represented by the length. Similarly, the 'land' between the pits along a track is from 0.8 to 3.6 $\mu$m in length, depending on the number of ones or zeros required. The track-to-track spacing is 1.6 $\mu$m.

The pits and lands are both 'read' by a laser. The laser is focussed on to a pit or land area through the polycarbonate substrate, as shown in Fig. 7.2. The moulded data surface is coated with a reflective film and this reflective film is covered with a thick protective layer to prevent it from being scratched. In Fig. 7.2 the laser beam is focussed on to a pit, and the scattered light is shown from the edges of the pit. This scattered light reduces the reflectivity and the strength of its signal received by the photodetector. When the laser is focussed on the lands, however, the reflecting layer ensures that the reflectivity is very high and so pit and land can be differentiated with ease.

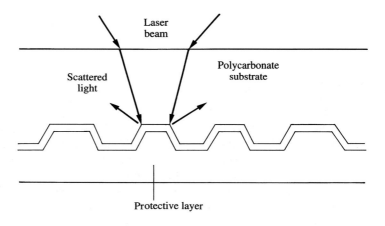

**Fig. 7.2** When the laser beam is focused on to the pit of the disc track, light scattering occurs resulting in a low level signal.

Figure 7.3 shows how the data pits on a CD-ROM are divided up into sectors in a continuous spiral. The labelling and the coding of these sectors will be given in detail below. Each sector has 2352 Bytes of information.

**Fig. 7.3** CD-ROM spiral of pits with identical sectors of 2352 bytes. In practice the sectors do not have to be separated so this saves valuable space.

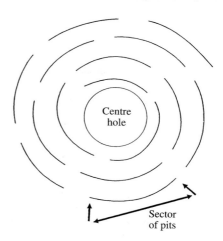

There are over 200 000 sectors in a CD-ROM spiral on one 120 mm diameter disc.

## 7.2 CD-ROM standards

Strong CD-ROM standards were adopted by the CD-ROM manufacturers in 1986 and 1988. In December 1986 the ECMA–119 European standard for volume and file structure for CD-ROM information interchange was issued which covers the software requirements. This was subsequently adopted by ISO as International Standard ISO 9660.

The specification of the disk itself was contained in the 'Yellow Book' which Philips and Sony gave to the licensees. In 1987 ECMA were asked to produce a standard based on the Yellow Book. In June, 1988 the ECMA–130 European standard was approved.

### 7.2.1 The ECMA–130 media standard for 120 mm CD-ROM discs

The disc standard is 45 pages long, so only selected parts of the standard which relate to media will be given here. (Free copies of the standard can be obtained from ECMA at 114 Rue du Rhone, 1204 Geneva, Switzerland.)

The *testing requirements* are specified as follows:

| | |
|---|---|
| Wavelength | 780 nm ± 10 nm |
| Polarization | Circular |
| Numerical aperture | 0.45 ± 0.01 |
| Intensity at the aperture stop of the objective lens | larger than 50 % of the maximum intensity value |
| Temperature | 23°C ± 2°C |
| Relative humidity | 45 % to 75 % |
| Atmospheric pressure | 96 KPa ± 10 Kpa |
| Conditioning before testing | 24 hours minimum |

For special environments, humidity testing is restricted to 45 to 55 %. No condensation on the disc is permitted.

Where applicable the disc is held between two consecutive rings for testing, having an inner diameter of at least 29 mm and an outer diameter of at most 31 mm holding the disc with a force in the range 1 N to 2 N.

The *operating environment* in which the disc has to be conditioned for at least 24 hours before operating:

| | |
|---|---|
| Temperature | −40°C to +70°C |
| Relative humidity | 10 % to 95 % |
| Absolute humidity | 0.1 gm$^{-3}$ to 60.0 gm$^{-3}$ |
| Sudden change of temperature | 50°C max. |
| Sudden change of relative humidity | 30 % max. |

No condensation on the disc is permitted.

*Storage environments* are also specified:

| | |
|---|---|
| Temperature | −10°C to 50°C |
| Short-term temperature | −5°C to 50°C |
| Relative humidity | 10 % to 90 % |
| Short-term humidity | 5 % to 90 % |
| Wet bulb temperature | 29°C maximum |
| Atmospheric pressure | 75 KPa to 105 KPa |

*Short-Term Storage* is defined as a maximum of 14 days. Regarding *physical properties*, the weight of the disc must be within 14 g and 33 g, so both polycarbonate and glass can in theory be used. In practice, only polycarbonate is specified or listed in any detail as follows:

| | |
|---|---|
| Disc outside diameter | 120 mm ± 0.3 mm |
| Hole diameter | 15.0 mm + 0.1 mm |
| | −0.0 mm |
| Concentricity | 0.2 mm |
| Thickness | 1.2 mm ± 0.1 mm |
| Index of refraction | 1.5 ± 0.10 |
| Birefringence (vertical) | 100 nm max. |
| Thickness of reflected layer | 55 nm ± 10 nm |
| Width of data pit | 0.50 ± 0.05 $\mu$m |
| Depth | 0.01 ± 0.02 $\mu$m |
| Length Range | 0.83 to 3.56 $\mu$m ± 0.05 $\mu$m |
| Spiral track separation | 1.6 ± 0.1 $\mu$m |
| Protective film thickness | 5 to 10 $\mu$m |
| Tampon printing | 3 $\mu$m thick |
| Screen printing | 9 $\mu$m thick |

## 7.2.2 Sectors of a digital CD-ROM data track

The digital data that is recorded in the pit and land lengths along an information track is represented by 8-bit bytes and grouped into 2352-byte sectors in the spiral already described above. The sector or block structure is defined in Table 6.1.

**Table 6.1**
**The ECMA–130 media standard**

| Designation | Size | Byte Number |
|---|---|---|
| Synchronisation | 12 Bytes | 0–11 |
| Header | 4 Bytes | 12–15 |
| User Data | 2,048 Bytes | 16–2063 |
| Error Detection | 4 Bytes | 2064–2067 |
| Space | 8 Bytes | 2068–2075 |
| Error Correction | 276 Bytes | 2076–2351 |

In practice, sectors are not physically separated, as was shown in Figure 7.3. The 12-Byte synchronization code at the beginning of each sector together with the number of bytes is used to identify the start of a new sector. Header data identify the sector uniquely in the first three bytes which contain the sector address. The fourth byte in the header is the mode byte. If it is Mode 1, this means that the user data are protected by error data code (EDC), error correction codes (ECC), and Cross Interleaved Reed-Solomon Code (CIRC). If it is set to Mode 2, then all the bytes in positions 16 to 2351 are user data bytes and the user data are protected by CIRC only.

The error detection code, EDC, is 4 bytes long and is a 32-bit CIRC, applied on bytes 0 to 2063. The error correction section consists of the last 276 bytes of the sector. Error correction will be discussed in more detail below after considering disc manufacturing.

## 7.3 Processing the CD-ROM disc

Before pre-mastering, stamper production, and disc replication are discussed the whole disc manufacturing process will be summarized. Figure 7.4 shows the **mastering process** for glass discs and this includes the pre-master stage. Pre-mastering involves arranging the data files into the ECMA–119 format and then recording with sector labelling and error correction coding on to the master tape. This master tape then drives the laser beam recorder so that the correct pit pattern is exposed on the disc.

**Fig. 7.4** Summary of the glass-based CD-ROM mastering process steps including pre-mastering.

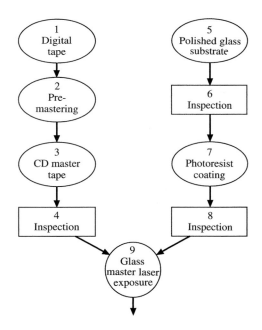

Figure 7.5 illustrates the **stamper production**. After the master has been exposed the pits are washed out, dried and inspected. Silver is evaporated or sputtered on to the glass master to produce a positive. In fact, normally two positive masters are made so that one can be kept for safety. One of these masters is then electroplated with nickel to form a negative nickel 'father'. The nickel is peeled from the master and the glass is then re-used to make another master. As many as ten positive nickel 'mothers' can then be made from this nickel father disc. From one mother as many as 100 negative nickel plated stampers can be made.

**Fig. 7.5** The stamper production process steps for nickel-plated CD-ROM masters.

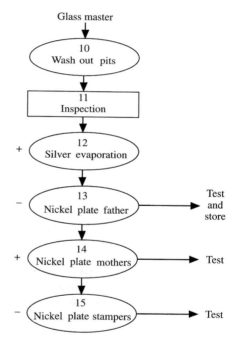

Glass master

10
Wash out pits

11
Inspection

+ 12
Silver evaporation

− 13
Nickel plate father → Test and store

+ 14
Nickel plate mothers → Test

− 15
Nickel plate stampers → Test

Then the **replication process** represents the final stage, as Fig. 7.6 shows. After polishing the back of the stamper it is placed in the injection mould. The polycarbonate discs are then replicated in the mould, and they are positives just like the master. Up to 10 000 discs can be moulded from one stamper. After moulding the discs the data surface is sputtered with aluminium, protected with lacquer which is cured with ultra-violet light, and then, after printing and drying the label on the lacquer, the disc substrate and data surface are inspected with a laser for defects. At the end of the process the discs which pass the inspection test are packaged in a jewel box or a cartridge.

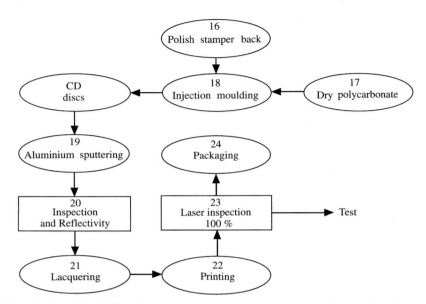

**Fig. 7.6** The CD-ROM replication process for producing polycarbonate discs from nickel-plated stampers.

## 7.4 Pre-mastering

Pre-mastering may be carried out by a CD-ROM publisher, software house or consultant. During pre-mastering the files containing the data and/or the directories are put into the ECMA–119 (or High Sierra) format.

The ECMA–119 standard can be obtained from the address given above (7.2.1).

The standard specifies:

(1)  the attributes of the volume and the descriptors recorded on it where 'volume' is a dismountable CD-ROM and 'descriptor' is a structure containing descriptive information about a volume or a file;
(2)  the relationship among volumes of a volume set;
(3)  the placement of files;
(4)  the attributes of the files;
(5)  record structures intended for use in the input or output data streams of an application programme, when such data streams are required to be

organized as sets of records where 'record' is a sequence of bytes treated as a unit of information;

(6)   three nested levels of medium interchange;

(7)   two nested levels of implementation where 'implementation' is a set of processes which enable an information processing system to behave as an originating system, or as a receiving system, or as both types of system;

(8)   requirements for the processes which are provided within information processing systems, to enable information to be interchanged between different systems, utilising recorded CD-ROM as the medium of interchange, for which purpose it specifies the functions to be provided within systems which are intended to originate or receive CD-ROM which conform to this ECMA standard.

The final stage of pre-mastering is to place the correctly formatted data, or compressed digitised audio, or compressed digitised video on to a 9-track tape, WORM disc or hard Winchester disc.

## 7.5 Mastering

The pre-mastered media then have the following added:

- error data code;
- error correction codes;
- CIRC code;
- channel encoding (bits turned to binary signal).

Once the coding has been added to every sector the CD-ROM master tape is prepared. This master tape drives the signal processing and the system controller of the laser disc recorder. The laser disc recorder exposes the photoresist in the spiral pattern and coding required. After washing out the exposed pits, the glass master is coated on both the photoresist side and the back side with silver.

## 7.6 Stamper production

The father, mother, and the stamper, shown in Figure 7.5, are all produced in the same nickel electroplating bath.

High purity water is required. Filtration by reverse osmosis and particle filtration to less than 0.2 $\mu$m eliminates most particles, bacteria and algae contamination. This high grade purity water has an extremely low ion content, giving conductivity less than 0.1 $\mu$S cm$^{-1}$. The water plant also has a water softener and an active carbon pre-treatment filter.

The nickel electroplating is started very slowly since the pit formation is the most critical part of the process. It is then speeded up so that a thick plate of 500 $\mu$m of nickel is produced in about six hours. During plating the current in the bath is programmable up to 200 A. A nickel clamp secures the glass master, nickel father or nickel mother which is in contact with the negative electrode.

During plating the large basket anode repels the positive nickel ions towards the cathode, so that nickel deposits uniformly on the master, father, or mother. The shape of the anode basket is important. Also critical for the electroplating are:

- temperature control to $\pm 1°C$;
- pH to $\pm 0.1$;
- filtering to submicron particles;
- the movement of the electrolyte.

After plating, thorough rinsing with the high purity water is essential.

Just prior to beginning the plating a conductive release layer is sprayed on to the master, father, or mother. This enables the nickel plate replica to be removed from the master, father, or mother.

Before placing the stamper in the mould for replication the back is polished either by hand or in a polishing machine. A final clean and inspection is carried out to ensure that there are a minimum number of defects on the critical front surface.

After the stamper has been polished on the back side a hole has to be punched accurately in the centre and the outer edge trimmed off.

## Question

If the stamper is a negative and the glass master is a positive why can only one be tested on a CD drive tester?

## Answer

Only the glass master can be tested, since it has pits in it that are identical to the polycarbonate CD that is replicated from the stamper. Both the master and the replicated polycarbonate disc are called 'positives'. The nickel-plated father that is replicated on the master as a 'negative' has bumps instead of pits. (The nickel mother that is plated on the father is a 'positive' and has pits, but the nickel stamper that is plated on the mother has bumps and is a negative.)

# 7.7 Replication

Figure 7.6 shows that the principal processes involved in replication are:

- injection moulding;
- aluminium sputtering;
- lacquering;
- printing;

- laser inspection;
- packaging.

Each one of these will now be described.

## 7.7.1 Injection moulding of polycarbonate compact discs

Polycarbonate granules are dried for four hours at 120°C to remove absorbed water. After this the dry granules are fed automatically, without exposing to the atmosphere, into the hopper of the plasticizing unit. Figure 7.7 shows a schematic diagram of the hopper and the rest of the plasticizing unit. The small screw on the right ensures that the granules feed into the central part at a uniform rate. The main plasticizer screw heats and melts the polycarbonate by mechanical shearing force. The temperature at the end of the main screw must be very carefully controlled.

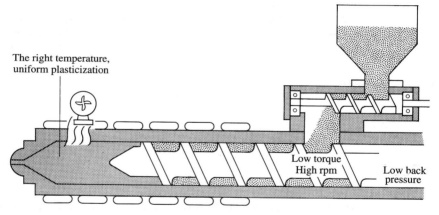

The right temperature, uniform plasticization

Low torque High rpm

Low back pressure

**Fig. 7.7** The polycarbonate feed system and plasticizing unit preceding disc moulding process.

In the high quality moulding machines seven process variables are controlled:

- screw revolution;
- back pressure;
- injection speed;
- holding pressure;
- barrel temperature;
- oil temperature;
- clamping force.

A closed loop control system that has been very successfully used in the compact disc industry is shown in Fig. 7.8.

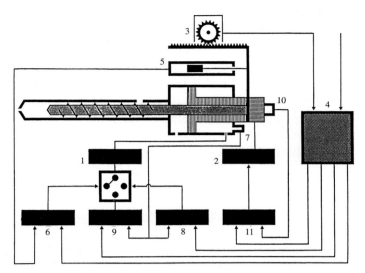

**Fig. 7.8** The closed loop control system for the injection moulding plasticizing unit manufactured by Netstal. (1) Servo valve which generates and maintains the programmed back pressure, injection speeds and holding pressure; (2) servo valve that controls the screw revolutions; (3) digital stroke measuring device with resolution of 0.01 mm; (4) computer programme unit; (5) linear velocity transducer; (6) injection speed controller; (7) transducer measuring the hydraulic pressure in the injection cylinder; (8) back pressure controller; (9) holding pressure controller; (10) tachogenerator measuring actual screw revolutions; (11) screw revolution controller;

The mould clamping unit is fully hydraulic and ensures the parallel travel of the mould platterns, and thus low mould wear. It also has fast opening and closing speeds which are very reproduceable, and the clamping force is constant.

It only takes six seconds to complete the moulding of one compact disc. The outer diameter and inner hole diameter are moulded to the required tolerance so no further finishing is required.

## 7.7.2 Reflective aluminium coating

After solidification in the mould the polycarbonate disc is removed by a small robot and transferred to the vacuum sputter coater. During this rapid transfer the disc remains warm and this prevents moisture condensing on it and helps the adhesion of the aluminium metal coat. There are four requirements for this metal film:

- high reflectivity so that the ECMA standard can be achieved (Section 7.2.1 above);

- good adhesion to the polycarbonate;
- low water content;
- no pin holes.

High reflectivity is easily achieved by sputtering aluminium from an aluminium target by bombarding it with argon. (Sputtering will be more fully described in Chapter 8.)

Adhesion is better with sputtered aluminium than with conventional vacuum evaporation for two reasons. First, as it is a higher energy process the aluminium atoms stick firmly to the surface, whereas with evaporation the lower energy aluminium atoms diffuse around the surface and form islands which have a low adhesion. Secondly, argon ions bombard the surface of the polycarbonate during the sputtering process to give some cleaning, which further improves the adhesion.

Water, which produces corrosion of the aluminium, is also lower in concentration with sputtering, since the vacuum pressure is lower.

Pin holes are very low with sputtering because the superior adhesion of the aluminium causes a very dense film to be formed. Upward sputtering from the target to the disc, or vertical targets and disc prevents debris falling down from the target on to the disc. Dust on the polycarbonate must also be removed by a nitrogen gas blow-clean just prior to placing the substrate in the sputter chamber. Dust causes pin-holes and can result in corrosion.

Sputter rates are typically 25 nm sec$^{-1}$ with 7 kW of power applied to the target cathode. This means that one disc with a metal film thickness of 55 nm takes just over two seconds for the aluminium to be deposited. Each target can produce 100 000 coatings.

## 7.7.3 Lacquering

A colourless acrylate lacquer is deposited over the aluminium to protect it. It is deposited on to the central portion of the disc and then spun at 4000 revolutions per minute for about 3 s. It is hardened under a UV lamp for another three seconds.

## 7.7.4 Label printing

Printing on to the lacquered CD-ROM surface can be accomplished with either of two techniques:

(1)   tampon offset or transfer printing;
(2)   screen printing

(1)   There are three key elements in tampon printing: the pad, the ink and the printing plate. The pad is used to transfer the image from the printing plate to the CD-ROM. It is made of an elastic silicone mixture. Special inks

have been developed by the printing machine suppliers for CD-ROM. The printing plate carries the printing image and is inked just before the pad contacts it and prints the labels on to the CD-ROM by ink transfer.

(**2**)    Screen printing involves printing through a nylon mesh screen. It results in a greater quantity of ink being deposited on to a disc, and the UV curing afterwards results in a very high gloss. As with Tampon offset printing, special quick-drying inks have been developed.

## 7.7.5  Laser inspection

A helium–neon gas laser red light inspection system is used to test every disc, after the printing process is complete. The diagram in Fig. 7.9 shows the optical system that is used. The multi-faceted mirror wheel produces a scanning beam on the disc which is 30 $\mu$m in radial width and 150 $\mu$m in tangential length. The tangential defect length is the length of a defect along the information spiral of pits. The disc rotates once and the complete disc scan is less than one second. The laser beam scans along the radius from the middle to the outer edge at a speed of about 180 metres per second and at a scan frequency of 3 kHz.

Two detectors, which can be photomultipliers or charged coupled devices, pick up the diffracted light beams. The zero-order detector picks up the directly reflected beam. The laser is plane polarised and so by using a

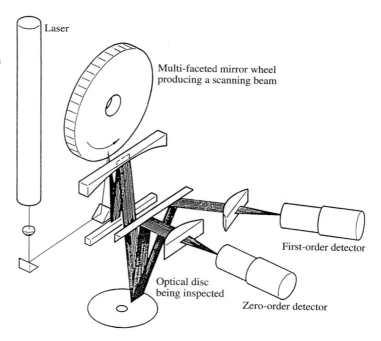

**Fig. 7.9** The optical inspection system for a CD-ROM disc as manufactured by Erwin Sick.

Laser

Multi-faceted mirror wheel producing a scanning beam

First-order detector

Optical disc being inspected

Zero-order detector

polarization filter in front of the zero-order detector, it can be made sensitive to local birefringence and detect small bubbles, inclusions and other defects in the polycarbonate substrate, which result in local stress. The first-order detector, which is used without a polarization filter, is very sensitive to metallisation defects, including pin holes, defects in pit structure and track separation. Up to ten defects can be detected with a minimum resolution of 0.3 $\mu$m. The detectors, after suitable amplification and signal processing, are linked up to a computer system so that a full analysis and continuous process monitoring can be carried out.

## 7.7.6 Packaging

A wide range of packaging is available for CD-ROM discs. For single discs the most common is the jewel box, which is identical to the plastic box that is used for CD-audio.

To avoid human contact with the disc, some companies have introduced a caddy from which the disc can only be unloaded inside the protected environment of the drive.

## 7.7.7 The replication production system

A simple yet very efficient production system which produces a high yield is shown in Fig. 7.10. The moulding subsystem's critical mould section on the

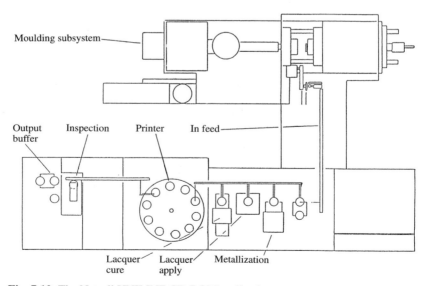

**Fig. 7.10** The Netsell UNILINE CD-ROM replication system.

right is protected with a vertical clean air shower, which extends over the other transport and process units right up to the output buffer. This means that when the compact disc is pulled out of the mould by a small robot, and travels down the 'In Feed' to the metallisation sputter station and thence on to the printer, the laser inspection station, and the output buffer store, the CD-ROM only sees class 100 clean air. (Class 100 means only 100 1 $\mu$m particles per cubic metre of air—all larger particles being excluded by air filtration.)

## 7.8 The CD-ROM data recording system

The CD-ROM data recording and reading system is shown in Fig. 7.11.

> The channel is defined as the route along which information may travel or be stored in a data processing system

The pre-master data are encoded to generate the master tape, which in turn creates the input binary synchronised channel bit stream, $B_i$, which controls the laser beam recorder, LBR, such that a signal is generated at A which switches the recording light beam on and off in order to expose the photoresist on the master disc, MD. The CD-ROM which has been replicated from the stamper, that originated from the master disc, is then READ by the optical head to give the read signal at $B_R$ in the CD-ROM drive. The read signal then has to be processed to give both the output signal, $B_0$, and the clock signal $C_L$. The output signal is finally decoded and passed via the interface to the computer.

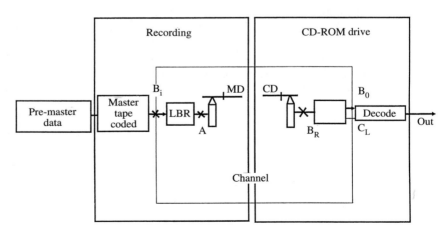

**Fig. 7.11** The CD-ROM data recording and reading system.

## 7.8.1 CD-ROM drive basics

Figure 7.12 shows the constituents of the CD-ROM drive:

- optical head;
- servo;
- spindle motor;
- digital signal processing;
- decoding and drive control;
- system control;
- interface.

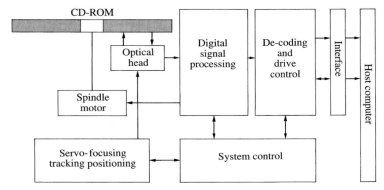

**Fig. 7.12** The basic constituents of the CD-ROM drive.

The optical head and the servo will be dealt with in detail below. The digital signal processing is shown schematically in Fig. 7.13. The optical head signal goes into the EFM, eight-to-fourteen demodulation (see below), which connects to the constant linear velocity (CLV) controller of the spindle motor. The EFM demodulation links to the de-interleave and interpolation to which the CIRC (cross interleaved Reed-Solomon error-correction code) error correction is added.

After error correction the signal goes on to the descrambling and drive control unit shown in Fig. 7.14. The interface to the host computer will be dealt with later.

## 7.8.2 Encoding and introduction to decoding

Figure 7.11 linked the recording and the drive, and showed the coding or encoding and decoding as just one step. In fact encoding and coding are more complex.

**Fig. 7.13** The digital signal processing in the CD-ROM drive.

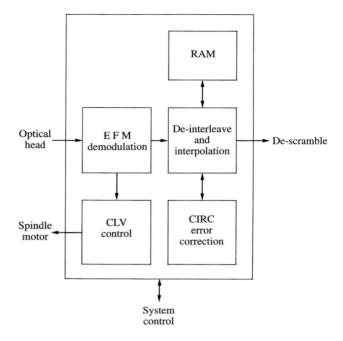

**Fig. 7.14** Descrambling and the drive control unit in the CD-ROM drive.

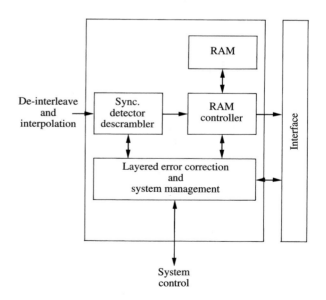

A summary diagram of the encoders and the decoders is shown in Fig. 7.15, Schouhamer (1991). The binary digital data from the pre-master tape has first an error correction code and then a recording code added to it. This provides the signal to control the laser disc recorder which produces the master. On the other side of the channel the recording code and the error correction code are then decoded before the data is sent to the interface and hence to the host computer.

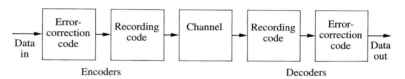

**Fig. 7.15** A summary of the encoders and the decoders in the CD-ROM data recording and reading systems using premaster data input.

On the encoding side the 16-bit digital signal has the CIRC (cross interleaved Reed-Solomon error-correction code) added. CIRC can handle both random and burst errors. The majority of the errors are random but burst errors can be caused by deep scratches and dirty fingerprint marks. The full encoding that is used in mastering to produce the signal that drives the laser beam recorder is shown in Fig. 7.16. (MUX stands for multiplexer.)

CIRC is a high quality error-correction system that was developed by Philips and Sony specifically for the compact disc. In the case of CD audio or CD-ROM it is the maximum length of a burst error which is critical.

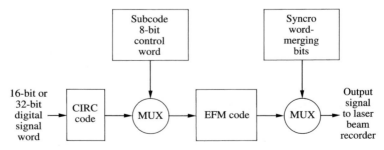

**Fig. 7.16** The encoding that is used in mastering to drive the laser beam recorder.

The BER, bit error rate, is defined as the number of errors received. The SIR, sample interpolation rate, is the number of data samples per unit time which have to be interpolated for given BER values. A low SIR means a superior system.

A specification for a CIRC system is shown in Table 7.1 (after Baert, 1988, p. 108).

**Table 7.1**
**Typical CIRC specifications for compact discs**

| Aspect | Specification |
|---|---|
| Maximum correctable burst error length | 4000 bits (i.e., 2.5 mm on the disc surface) |
| Maximum interpolatable burst error length | 12 300 bits (i.e., 7.7 mm on the disc surface) |
| Sample interpolation rate | One sample every 10 hours at a BER of $10^{-4}$ |
| | 1000 samples every minute at a BER of $10^{-3}$ |
| Undetected error samples | Less than one every 750 hours at a BER of $10^{-3}$ |
| | Negligible at a BER of $10^{-4}$ |
| Code rate | On average, four bits are recorded for every three data bits |

The CIRC encoder limits pit lengths to:

a minimum of 3T;
a maximum of 11T,

where T is the clock period of the system. For T = 231 ns and a speed of 1.2 ms$^{-1}$ the minimum and maximum pit lengths equate to:

$3T \equiv 0.833 \ \mu m$
$11T \equiv 3.05 \ \mu m$

## 7.8.3 The optical head

Now moving to the CD-ROM drive components, the first key part to understand is the optical head, since this is used to read the data from the spiral pit sequence on the CD-ROM disc.

A three-beam optical head is shown in Fig. 7.17 (after Baert, 1988, p. 135). The gallium arsenide laser light beam is incident upon a diffraction grating and this diffracts the light into a number of diffraction orders. In the figure only the zeroth order and the first two orders on either side of it are shown, since the higher orders are elliminated by the optics. These two-side first-order light spots are used for tracking accuracy, as will be seen below. After the grating, the three light spots strike a polarizing beam-splitting prism. Only the vertically plane-polarized (with the *E* vector vertical) laser light passes through to the collimation lens. The collimator produces a parallel beam which then strikes the quarter wave plate, which rotates the plane-polarized light through 45°. The three beams are then focussed on to the CD-ROM by a simple bi-convex lens. The lens is fitted into the fine-focus actuator, which will be described below.

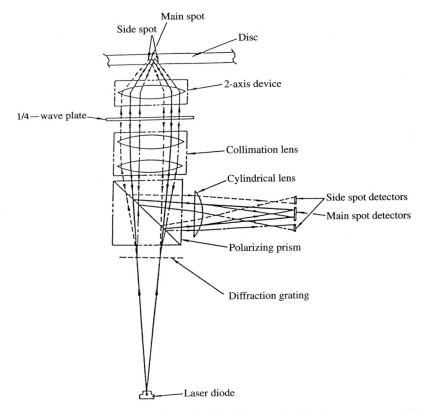

**Fig. 7.17** A three-beam CD-ROM optical head for reading the pit sequence on a CD-ROM disc.

After reflection from the aluminium reflective layer on the CD-ROM the beams return through the focus lens to the quarter wave plate, where the polarization is again rotated through 45° to make the beams horizontally polarised. The beams pass through the collimator again and are reflected by the prism beam splitter into a cylindrical lens, which focuses them on to the six detectors. Four detectors are used for focussing the zeroth order, and the two side spot detectors are used for tracking.

The fine-focus actuator is shown in Fig. 7.18 (after Baert, 1988, p. 138). It is a two-axis or two-dimensional device with a single focus coil and curved permanent magnets, as shown. The force generated by this coil and its magnets moves the lens up and down against some restraining wires and a counterbalance. How focussing is achieved will be dealt with below when the control systems are discussed. The two tracking coils have two permanent magnets which are positioned very close to them. A current in the left-hand side coil tilts the beam to the left, and a current in the right-hand side coil tilts it to the right.

## 7.8.4 Control systems in the CD-ROM drive

To explain the principles of the control systems in the CD-ROM drive only the focus control will be considered in detail. Figure (Bouwhuis, *et al*., 1985) shows the fine focus control system, and for simplicity the permanent

**Fig. 7.18** The fine-focus actuator showing focus coil and magnets.

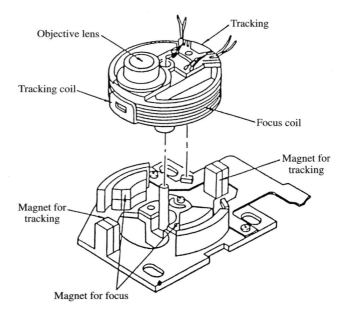

**Fig. 7.19** Close-up of a CD-ROM head showing lens and tracking gear.

magnets are not shown. The optical position detector will be discussed fully below. This detector, which supplies a voltage signal approximately proportional to focus error, is connected to the pre-amplifier, whose output goes to the compensation network for filtering. Finally, the output from the compensation network goes to the driver so that a current flows through the actuator coil, and this produces a vertical force on the objective lens. Figure 7.20 shows a closed loop or feedback control system for fine focus.

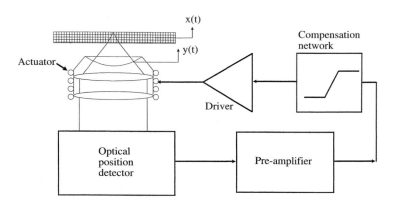

**Fig. 7.20** The fine focus control system with feedback.

The frequency response of a typical spring-supported actuator is shown in Fig. 7.21 (after Marchant, 1990). The actuator response in mm A$^{-1}$ is shown. The actuator response is constant at low frequencies because the spring forces dominate. After the principal resonance of the system the inertia dominates and the $w^{-2}$ slope prevails. However, at high frequencies above 1 kHz, undesirable parasitic resonances occur.

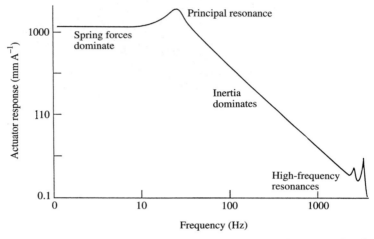

**Fig. 7.21** A typical frequency response of a spring–supported actuator showing the dominant $w^{-2}$ characteristic.

## 7.8.5 The detector system for focusing and tracking

The detector system for a three-beam optical head is shown in Fig. 7.22. The quadrant detector ABCD is used for focusing and the two side spot detectors are used for tracking (Baert, 1988, p. 147).

Figure 7.23(c) illustrates the correct tracking position with the central zeroth order spot perfectly centred on the pit of one of the CD-ROM tracks. The two-side first-order spots just clip the left-hand side and the right-hand side of the pits. Poor tracking or mis-tracking is shown in 7.23(a) and (b). The spot separation is approximately 20 $\mu$m.

**Fig. 7.22** The detector system for a three-beam optical head showing how the various error signals are obtained from the photodetectors of the CD optical pick-up. *RF*, radio frequency data signal; *TE*, tracking error signal; *FE*, focus error signal.

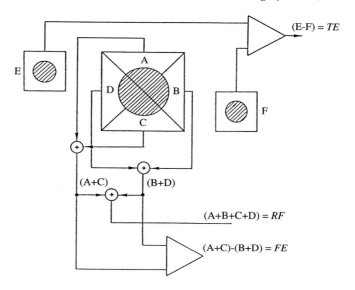

**Fig. 7.23** (a) and (b) Mistracking and (c) correct tracking on the CD-ROM pits for the three-beam optical head.

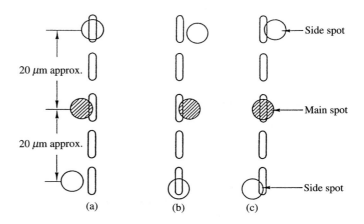

Returning to Fig. 7.22, the two-side spot detectors E and F give the **tracking error** signal *TE*:

(7.1)
$$TE = E - F$$

When the signals on E and F are equal there is no tracking error.

The **focus error**, *FE*, is given by subtracting the summation of opposing quadrants of the quadrant detectors:

(7.2)
$$FE = (A+C) - (B+D)$$

Perfect focusing is illustrated in Fig. 7.24 with *FE* = 0, since the circular spot gives an equal signal from each quadrant detector (Beart, 1988, p. 144). Poor focusing gives an elliptical spot and *FE* no longer equals zero.

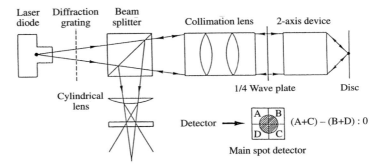

**Fig. 7.24** The main spot detector with a perfectly focused light spot and the components of the optical head with a light-beam ray diagram.

Looking again at Fig. 7.22, the radio frequency, *RF*, current signal that contains the digital data information is given by:

(7.3)
$$RF = (A + B + C + D)$$

This r.f. signal is the signal from the 'ones' and the 'zeros' on the disc. As soon as *RF* reaches the threshold level shown in Fig. 7.25(b), the focus servo is activated and controls the fine focus actuator until the zero *FE* signal in Fig. 7.25(a) is achieved. At this zero *FE* value, *RF* reaches a maximum.

## 7.8.8 The coarse actuator and disc motor servos

Tracking error signals are used to control the sled or the coarse actuator. The coarse actuator moves the complete optical head. When the tracking error signal reaches a certain threshold the sled is activated and moves until the same threshold is again reached.

**Fig. 7.25** The focus error (*FE*) signal (a) and the radio frequency (*RF*) output signal (b) illustrating how the focus servo circuit is only activated at signals above the r.f. threshold level.

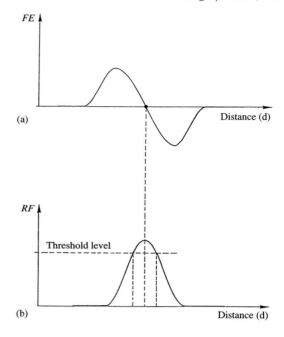

The disc motor servo, on the other hand, is activated by the synchronisation code signals on each sector. The synchronisation frequency is compared with a crystal oscillator in a phase comparator, and the motor speed corresponds to the phase or frequency difference. A constant linear velocity is required so that the disc gradually increases its speed as the reading optical head moves from the inner to the outer disc tracks near the outside edge.

## 7.8.9 Signal processing

The simplified block diagram in Fig. 7.26 shows the key elements of signal processing (Baert, 1988, p. 151):

- r.f. amplification;
- wave shaping;
- synchronisation;
- EFM demodulation;
- error detection and correction.

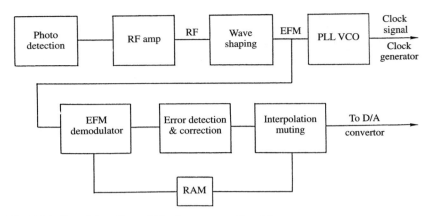

**Fig. 7.26** Signal decoding within the compact disc drive.

The r.f. signal A+B+C+D from the quadrant detector is a weak current signal. Consequently, this is converted to a voltage signal and then amplified. Figure 7.27 illustrates the r.f. amplifier circuit and shows where the 'eye pattern' data output signal can be observed (Baert, 1988, p. 149). IC402 includes the current to voltage conversion and IC403 corrects the offset voltage or asymmetrical signal and amplifies it.

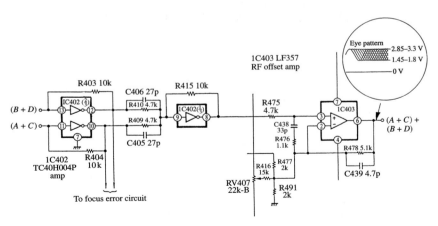

**Fig. 7.27** The r.f. amplification circuit.

The eye pattern is more clearly shown in Fig. 7.28. The amplitude of the signal $I_3$ corresponds to a pit 3 clock periods long and $I_{11}$ corresponds to a pit 11 clock periods long. The diamond eye should be as sharp and the amplitude as high as possible. The eye width is 0.3 $\mu$m. $I_3$ is the highest, 720 kHz, and $I_{11}$ is the lowest, 196 kHz, frequency. The top level of r.f. signal is $I_{Top}$ and this includes the zero offset or d.c. bias.

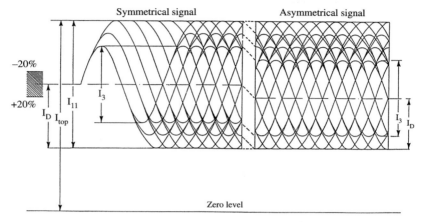

**Fig. 7.28** RF signal 'eye pattern' after amplification.

The information contained in the r.f. signal is extracted in the form of the positions of the crossings of the signal with a decision level, $I_D$, in the centre of the eye pattern. There are three 'eyes' between successive crossings of $I_3$ and $I_D$ corresponding to three clock periods.

The ECMA–130 standard defines the modulation amplitude as:

$$0.3 < \frac{I_3}{I_{Top}} < 0.7$$

$$\frac{I_{11}}{I_{Top}} > 0.6$$

and the symmetry of the R.F. signals with respect to $I_D$ as :

$$-20\% < \left( \frac{1}{2} \cdot \frac{I_D}{I_{11}} \right).100\% < +20\%$$

The **cross talk** is measured as the ratio between the r.f. signal amplitude with the beam halfway between two tracks and with the beam on track. This ratio has to be less than 0.5.

Now, returning to the block diagram in Fig. 7.26, a wave-shaping circuit is used to detect the offset zero-cross points of the eye pattern and convert them into a coded digital signal. The timing diagram of the r.f. signal before and after waveshaping (Baert, 1988, p. 150) is shown in Fig. 7.29. This shows the amplitude of the r.f. signal varying with pit length. Since the pit causes light scattering the loss in amplitude is greater the longer the pit, and similarly, the output increases to higher levels for larger gaps between the pits. The cross-over points of the eye pattern determine the switch from a '1' to a '0' in the digital signal.

After waveshaping, the signal is integrated in the feedback loop through a low pass filter to produce a d.c. voltage which is applied to the input, so that the correct 'slicing' for the eye pattern signal is obtained.

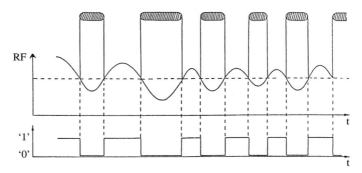

**Fig. 7.29** Timing diagram of the r.f. signal before and after waveshaping.

## 7.9 Outline specification for a CD-ROM drive

Toshiba provided the following outline specification for their XM-3201B drive. This uses a CD-ROM disc protected in a cartridge.

| | |
|---|---|
| Capacity (Mbytes) | 599 |
| User data per sector (bytes) | 2048 |
| Number of sectors per disc | 292.5k |
| Rotational Speed (rev min$^{-1}$) | 530–200 |
| Data Transfer Rate (kbits/sec) | 153.6 (Avge) 1400 (max) |
| Access Time (ms) (includes latency): | |
|     Track-to-track | 1 + Latency |
|     Average Random | 350 |
|     Maximum | 700 |

| Motor Start (ms) | 1000 |
| Interface system | SCSI |
| Mean-time-between-failures | 25 000 hours |
| Error rates: | |
|    Unrecoverable | 1 in $10^{12}$ bits |
|    Seek | 1 in $10^6$ bits |

## 7.10 Applications

The CD-ROM is developing very rapidly as multi-media applications arise in which data, audio and video are all combined, so this section will be restricted to a few of the data storage applications.

Medium to high volume of 100 to 100 000 copies is the area where CD-ROM publishing will dominate. Among the many possible applications the following are just a few where CD-ROM is already making an impact:

- dictionaries;
- legal directories;
- medical data;
- safety hazards information;
- software for business computers;
- educational courses or modules;
- industrial training;
- maps;
- manuals;
- marketing information;
- technical information;
- company directories;
- annual reports;
- a one-year issue of a newspaper;
- patents.

For large volumes the production costs for each CD-ROM are much lower than for a book. Hence, the cost per bit of information is much cheaper still, since one single CD-ROM disc stores over 200 000 A4 pages.

## References

Baert, L. (1988) *Digital Audio and Compact Disc Technology*, Heinemann, pp. 108–9.
Bouwhuis, G., *et al*. (1985) *Principles of Optical Disc Systems*, Adam Hilger, p. 137.
Marchant, A. B. (1990) *Optical Recording*, Addison-Wesley, p. 185.
Schouhamer, K. A. (1991) *Coding Techniques for Digital Recorders*, Prentice Hall, p. 3.

# Magneto-optic thin film production

**8**

## 8.1 Evaporation

Sputtering occurs when a high energy inert gas strikes a metal target and knocks out target atoms. Conventional evaporation however, is a much lower energy process as Fig. 8.1 shows (Takagi, 1982). The kinetic energy of the source materials that can be evaporated from raw material with a high current flowing through it or by using a scanning electron beam, varies from $1 \times 10^{-2}$ to 1 eV. This process of evaporation is used to produce finite quantities of pure material as magnetic thin films for data recording.

When the evaporated atoms condense on the substrate they tend to diffuse sideways along the surface and hence pin holes are often formed, particularly for very thin films of less than 10 nm and wherever there is a large defect in the surface or if contamination is sticking to the surface.

**Fig. 8.1** Kinetic energy of the source materials used in various vacuum processes for the production of magnetic and magneto-optic thin films.

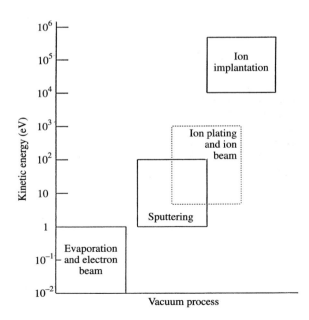

For magnetic and optical recording, these pin holes cause errors in the data and they can also lead to corrosion in the magneto-optic film since rare-earth transition metal magnetic materials used in the process are very reactive to atmosphere.

In addition, because conventional evaporated films are deposited at low energy they do not have good adhesion and also often change their composition appreciably.

It was these limitations to producing thin film magnetic media that urged researchers to find an alternative vacuum technique to evaporation. Sputtering with its higher energy for the source material is an attractive alternative. Magnetron sputtering which is described below is now established as a commercial process for the preparation of thin film media for Winchester disc drives and for optical media including CD-ROM, WORM and eraseable WREM.

## 8.2  Sputtering processes

The basic sputtering process is illustrated in Fig. 8.2. An argon ion is incident on the surface of the target raw material and this knocks out target atoms which are called sputtered atoms. Reflected ions and 'neutrals' or neutral argon ion atoms, and secondary electrons are also produced. Good adhesion onto the substrate depends mainly on the cleanliness of the source material since nucleation occurs at many different sites on the substrate. There is no side-ways diffusion. Indeed, for higher ion energies the sputtered atoms actually 'alloy' with the substrate in certain cases to give a graded transition from the substrate to the thin film during sputtering and so reduce strain at the film/substrate interface.

**Fig. 8.2** A positive argon ion impacting with the target surface during the basic sputtering process to produce thin film layers of pure target (source) material under vacuum.

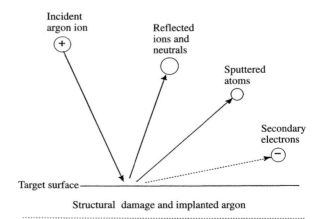

The target (source) composition is therefore faithfully reproduced in the deposited (substrate) film if the target is a perfect alloy. However, if there are any other alloy phases present in the target these 'impurities' will sputter at different energy levels and so the composition of the coating on the substrate may differ considerably.

## 8.2.1 DC and r.f. Sputtering

In d.c. sputtering a voltage is applied between two electrodes after argon to a pressure of 1 to 100 millibars has been admitted to the evacuated chamber. As shown in Fig. 8.3 a negative bias is applied to the target (source material) or cathode and the substrate or anode is biased positive. Consequently, the electrons are accelerated between the electrodes. These electrons ionize the neutral argon atoms and the resultant positive argon ions are then attracted to the negatively biased target and hence sputtering is achieved. The argon ionization plasma glows purple and covers a wide region between the electrodes when a stable situation has been achieved and ionization is occuring at a reasonably constant rate.

**Fig. 8.3** DC diode sputtering process

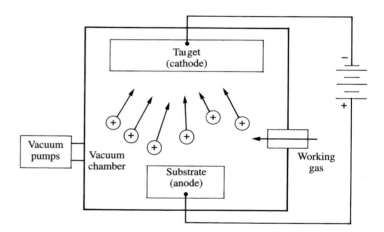

Returning to Fig. 8.2, it should be pointed out that the secondary electrons can enhance the ionization rate. Mention should also be made of the fact that neutral argon atoms can also be incorporated into the film and this can affect the stress in the film.

RF sputtering with r.f. biasing at a frequency of 13.65 MHz is shown in Fig. 8.4. Careful matching of the RF network is required. This type of biasing enables non-conducting oxides and nitrides to be used as cathodes or targets. The sputtering yield is low but high quality films can be obtained particularly if the target material is chemically cleaned first.

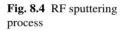

**Fig. 8.4** RF sputtering
process

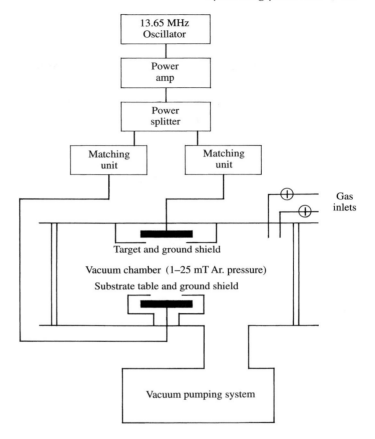

## 8.2.2 Magnetron sputtering

The magnetron source was developed by Penfold and Thornton. It uses a magnetic field to trap electrons close to the sputtering target. Figure 8.5 shows the operation of the DC Magnetron. The magnetic field causes free electrons to hop round a 'race-track' restricted to the vicinity of the cathode. This electron confinement enhances the ionization rate of the argon atoms close to the cathode and produces a very high rate of sputtering. RF magnetrons are not as efficient but they still give an increased yield over the standard RF bias technique.

**Question**

What actually happens at a magnetron target, and what is the difference between d.c. and r.f. bias?

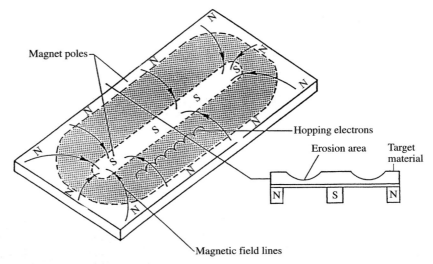

**Fig. 8.5** A magnetron method of sputtering showing the hopping electrons of the target (source) being confined to a 'race-track' region by the magnetic fields from the magnets under the target.

**Answer**

In the case of DC bias the magnetron is negatively biased and is the cathode. Positive argon ions are attracted from a plasma between the cathode and the anode on which the substrate is fixed. These positive argon ions sputter or knock off atoms from the heated target and generate secondary electrons in addition to those primary electrons generated as current between the two electrodes. The magnetic field shown in Fig. 8.5 between the permanent magnets mounted under the target confines the electrons so that they hop round the target material as around a race-track (shaded). These electrons ionise the argon and create more positive argon ions. These additional positive ions are attracted to the negative cathode and in so doing sputter target material which then deposits on the substrate anode to create a thin film layer with magnetic properties.

With r.f. bias the target alternates between positive and negative bias since the current is alternating. During the negative bias it acts like a cathode and deposition on the substrate, which is also r.f.-biased, occurs. The r.f. bias on the substrate must match the target, that is, be positive when the target is negative. RF biasing is slower than d.c. since sputtering occurs for less than half the time. However, r.f. biasing prevents charge build-up on insulator targets like glass or ceramic.

### 8.2.3 Ion plating

In ion plating, the metal, compound or alloy is evaporated into the argon plasma. In this way the material to be deposited is ionized and deposited onto the substrate cathode at a very high energy. Excellent adhesion is achieved both due to cleanliness and the high energy of the depositing material ions. Alloying with the substrate can occur and for certain applications this is highly desirable. Fig. 8.6 shows the equipment used (Mattox, 1964) for ion plating.

**Fig. 8.6** An early ion plating equipment used for thin film production (Mattox 1964).

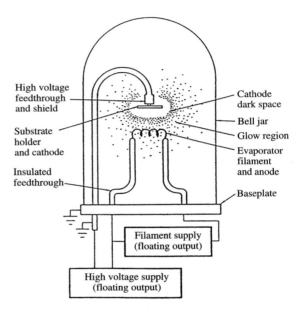

In **sputter-ion plating** a magnetron replaces the filament or electron beam heated source.

### 8.2.4 Ion beam deposition

An ion beam source from an ion beam gun produces ions of the material that is to be deposited. These ions can be isolated or clustered in a beam. Clusters which consist of about 1000 individual atoms loosely coupled together are generated by adiabatic expansion of the vapourized source material through a nozzle. A large percentage of these clusters are ionized by electron bombardment and are accelerated by an acceleration voltage field in the direction of the substrate. T. Takagi and colleagues have been particularly successful in depositing films with a high degree of crystallinity with this technique (Takagi, Yamada and Sasaki, 1976).

## 8.2.5 Argon ion beam source sputtering

An argon ion beam source is an excellent tool for sputtering. Figure 8.7 shows a Kaufman ion beam source sputtering equipment.

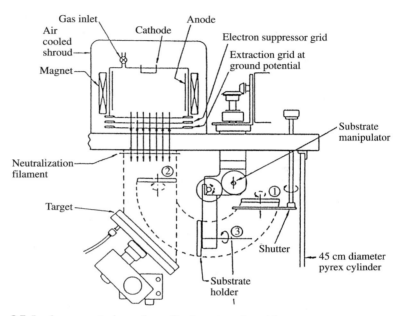

**Fig. 8.7** Ion beam sputtering using a Kaufman-type broad beam source.

The argon ions are extracted as a broad beam and when the substrate is in position 1 the target is 'cleaned' by the bombardment of the argon ions. Substrate sputter cleaning is achieved by moving the substrate to position 2. Finally, in substrate position 3, sputter deposition takes place: the argon ions sputter off the heated target (source) atoms which are then deposited on the substrate. This is a very clean deposition technique and yields very high quality optical films which have been in use for laser mirrors and optical filters for over 20 years. Surprisingly, magnetic and magneto-optic films have not been regularly deposited by this method until quite recently.

# 8.3. Magnetic film deposition for Winchester hard discs

A schematic diagram of a CPA 9930 system is shown in Figure 8.8. This sputterer was originally developed for the semiconductor industry but it was one of the first machines to be used to sputter thin magnetic films on to Winchester disc substrates. Originally it was designed to coat only one side of 3.5, 5.25 or 8 inch discs in the same way as only one side of a silicon wafer was coated. Figure 8.8 shows the single-sided coater with four magnetrons sputtering down onto the substrates which move along a conveyor from the entry lock to the exit lock. The substrates are loaded onto pallets under a Class 100 clean air hood until a full batch is ready to be placed in the entry lock. The entry lock is then evacuated with a cryopump system to a vacuum of below $10^{-6}$ bar. On achieving this vacuum the lock is opened to the magnetron line and the first pallet is fed onto the conveyor belt. The linear movement of the substrates through the magetron line give a thin film thickness uniformity of ±5 % (With a stationary substrate a ±10 % variation occurs according to Drennan, 1985.)

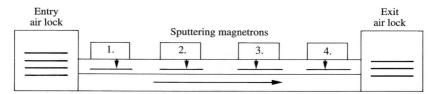

**Fig. 8.8** The CPA 9930 four-stage magnetron production sputterer for rigid disc, thin film, magnetic media. Substrates are moved automatically under vacuum from left to right.

To complete the hard disc coating with the sputterer shown in Fig. 8.8 all the discs are turned over and put through the line again. An obvious development was therefore to coat both sides at the same time in a horizontal or a vertical coater. Vertical coaters can coat up to 100–200 discs per hour as supplied by Leybold or Balzers. Again the discs are loaded manually or preferably with a robot on to a pallet mounted on a conveyor belt. The vertical configuration is preferred since debris from the magnetron source will not land on the surface of the disc. The vertical sputtering machine is built in a modular fashion so that additional sputtering modules can be inserted if extra coatings are required.

For hard disc media processes that are currently available there are two options that dominate the market. In the partially-sputtered disc shown in Fig. 8.9 the sputtered magnetic alloy is sputtered on to a plated nickel phosphorous alloy layer and then a carbon overcoat is sputtered on to protect the magnetic film. The nickel plating is 10 to 15 $\mu$m thick and has to be polished and textured before the magnetic coating is applied. Texturing produces a grooved surface which assists the magnetic head aerodynamics. The magnetic coating is only 25 to 50 nm thick and the carbon coating can go down to 10 nm.

**Fig. 8.9** A partially-sputtered rigid disc showing two thin films in addition to nickel plating.

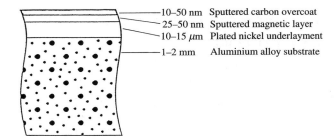

10–50 nm   Sputtered carbon overcoat
25–50 nm   Sputtered magnetic layer
10–15 $\mu$m   Plated nickel underlayment
1–2 mm   Aluminium alloy substrate

More recently a fully-sputtered disc has been available. Figure 8.10 shows the cross-section of this type of media. A sputtered chromium 'underlayment' is used to seed the magnetic layer onto the aluminium alloy substrate. The thickness of the chromium layer required depends on the magnetic alloy and domain orientation required.

**Fig. 8.10** A fully-sputtered disc cross-section. The substrate is an aluminium alloy or glass.

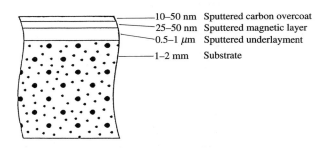

10–50 nm   Sputtered carbon overcoat
25–50 nm   Sputtered magnetic layer
0.5–1 $\mu$m   Sputtered underlayment
1–2 mm   Substrate

The most promising material which is still in the research stage is cobalt-platinum-chromium Yogi *et al.* (1990). A 1 Gb per square inch areal density longitudinal media has been achieved using two approaches:

(a)   High bit per inch (BPI): 160K BPI × 6.4K TPI
(b)   High track per inch (TPI): 120K BPI × 8.5K TPI

The respective thicknesses of the layers in the media was 12.5/23/100 nm for C/CoPtCr/Cr sputtered films. Track widths of 4 microns were achieved.

Another alternative alloy material is cobalt-chromium-tantalum on a chromium underlay. This has been prepared by DC magnetron, Pressesky *et al.* (1990) RF diode sputtering, Hau, Sivertson and Judy, (1990) and facing target sputtering, Kawanbe, Hasegawa and Nase (1990). Coercivities of 1650 Oersted have been achieved.

# 8.4 Sputtered films for commercial optical discs

## 8.4.1 CD-ROM

In this case sputtering of only an aluminium layer is required and the parameters for this are as follows using a Balzers CDI 800 sputtering machine:

Substrate: 120 mm diameter, 1.2 mm thick, polycarbonate;
Film: aluminium, 55 nm, reflectivity greater than 87 %;
Capacity: 720 discs per hour.

To prevent contamination, lacquering is carried out in clean air conditions immediately after the sputtering is complete and the high temperature prevents water absorption and assists in the bonding to the lacquer.

## 8.4.2 WORM (Write-once-read-many)

There are many different technologies for WORM (see Chapter 9):

(a)  Mirror
(b)  Hole or Pit
(c)  Phase change
(d)  Bubble

The majority of these use just one sputtered layer on a polycarbonate substrate. Currently, the most reliable is based on platinum on a specially structured surface produced by Plasmon Data Ltd. On heating the sputtered film with a laser the structure is relieved and a 1 μm diameter mirror is formed to record the '1' or '0' bit of digital information.

Nevertheless, the pit type dominates the market and a range of sputtered layers has been used from TeSe, TeOx, and TeC; also dye-based pit versions exist. Mirror versions have been produced made from platinum, and bubble technology involves two layers of proprietary plastic. A cross-section of a tellurium alloy WORM disc is shown in Fig. 8.11.

## 8.4.3 WREM (Write-read-erase-many)

Currently, sputtered magneto-optic films are used in commercially-available media. The four layer structure of these films is as shown in Fig. 8.12. The substrate which is either polycarbonate or glass is first coated with 20 nm of aluminium nitride either by reactive sputtering using an aluminium magnetron target or by r.f. magnetron sputtering with an aluminium nitride target. The magneto-optic layer is terbium-iron-cobalt and is 50 nm thick. This is generally d.c. sputtered from an alloy target. The composition of the alloy determines the thermo-magneto-optic properties. Aluminium nitride 20 nm thick is again used for the third coating for protection of the magneto-

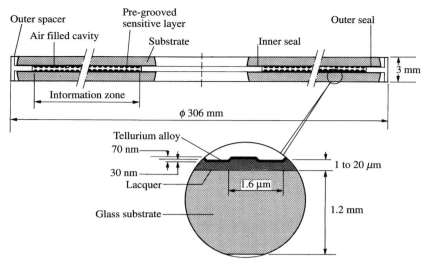

**Fig. 8.11** A typical WORM disc cross-section.

optic films and as an anti-reflective coating. This is followed by a reflective aluminium layer of at least 10 nm thickness. All four layers are deposited in an in-line sputterer and sputtering rates of 360 discs per hour have been achieved.

**Fig. 8.12** The four sputtered layers on a magneto-optic WREM disc.

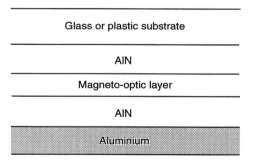

In commercial recording systems, carrier-to-noise ratios of well in excess of the 47 dB standard level have been achieved. On a 5.25 inch disc recording capacities of over 1 Gbyte have been reported with magnetron sputtered media (Philips 1989 press release).

The main worry with current opto-magnetic media is degradation due to oxygen or water reaction with the rare earth constituent of the magneto-optic alloy film. Figure 8.13 shows the degradation of the reflectivity with time in an atmosphere of 80 % RH of various thin film technologies. Tellurium-carbon WORM media shows much less degradation than the rare earth based layers. (The sputtered layers used to obtain the data were all unprotected so this was an accelerated life test.)

**Fig. 8.13** The reflectance change of WORM and magneto-optic films as they are exposed to environmental conditions of 70°C and 85% humidity. There is no protective coating on any of the films. (Courtesy: Toshiba.)

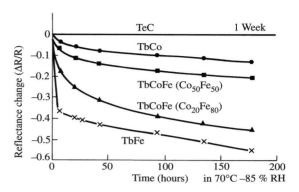

Of all of the possible **protective layers**, aluminium nitride has been found to be the most satisfactory. Media with this protective coating has now been in permanent use for several years and has shown little sign of degradation. Research is in progress to find the next generation of single magneto-optic layer materials. Of these, various ferrites are very stable and could be used without a protective coating in some applications.

Multi-layer coatings are also attracting considerable interest. Cobalt-Platinum with a total of 25 multi-layers including 0.26 nm cobalt and 1.26 nm platinum, prepared with d.c. magnetron sputtering in krypton has been used to produce media with a carrier-to-noise ratio in the region of 50 dB (Carcia and Zeper, 1990).

DC magnetron sputtering has also been used to produce multi-layers of terbium with iron-cobalt, Che, *et al.* (1990) and iron-cobalt-titanium, Toki, *et al.* (1990). In both cases media with a carrier-to-noise ratios of greater than 47 dB have been produced.

# 8.5. Conclusion

Sputtered magnetic and magneto-optic media are both commercially available. Magneto-optic media was only available from 1988. The search for the second generation magneto-optic media in single or multi-layer form is now under way but sputtering will almost certainly be used to produce it.

Of all the sputtering technologies, d.c. magnetron is widely accepted as the best for Winchester rigid discs and all types of optical media.

# References

Carcia, P. F., Zeper, W. B. (1990) *IEEE Transactions on Magnetics*, **MAG-26**, pp. 1703.

Carter, G., and Colligon, J.S. (1968) *Ion Bombardment of Solids*, Elsevier.

Che, Y., Yang, C., and Shun, S. (1990) *IEEE Transactions on Magnetics*, **MAG-26**, p. 1712.

Drennan, G. A. (1985) *Hewlett-Packard Journal*, November.

Hau, Y., Sivertson, J. M., and Judy, J. H. (1990) *IEEE Transactions on Magnetics*, **MAG-26**, p. 1599.

Humphries, M., and Williams, E. W. (1989) *Quality*, p. 56, November.

Ishikawa, M., Terao, K., Hashimoto, M., Tani, N., Ota, Y., and Nakamura, K. (1990) *IEEE Transactions on Magnetics*, **MAG-26**, p. 1602.

Kawanabe, T., Hasegawa. K., and Nase, M. (1990) *IEEE Transactions on Magnetics*, **MAG-26**, p. 1593.

Laynovsky, W., (1975) *Research/Development*, 47–54, August.

Mattox, D. M. (1964) *Electrochemical Technology*, **2**, pp. 295–8.

Penfold, A. S., and Thornton, J. A., US Patents 3884793 (1975); 3995187 (1977); 4030996 (1977); 4031424 (1977); 4044353 (1977).

Pressesky, J. L., Lee, S. Y., Williams, D., Heiman, N., and Coughlin, T. (1990) *IEEE Transactions on Magnetics*, **MAG-26**, p. 1596.

Takagi, T. (1982) Role of ions in ion-based film formation, *Thin Solid Films*, **92**, pp. 1–17.

Takagi, T., Yamada, I., and Saski, A. (1976) *Thin Solid Films*, **39**, p. 207.

Thornton, J. A., and Hoffman, D. W. (1980) *27th National Symposium for the American Vacuum Society*, Detroit, October.

Toki, K., Kariyada, E., Shimada, M., Yarmoto, S., and Okado, O. (1990) *IEEE Transactions on Magnetics*, **MAG-26**, p. 1709.

Yogi, T., Tsang, C., Castillo, G., Gorman, G. L., Ju, K., Nguyen, T. (1990) *IEEE Transactions on Magnetics*, **MAG-26**, p. 2271.

# 9

# WORM media

Write-once-read-many (WORM) optical media was slow to gain in popularity when it was initially developed. Between 1984 and 1990 there were no standards and there were at least five different technologies. Most of these technologies required unique drives and were not interchangeable. In addition, a wide range of disc sizes existed, from 12 inch down to 3.5 inch.

The situation has improved with the introduction of standards and concentration on archival storage. In spite of this, document image processing systems tend to use a magneto-optic media because it is erasable and therefore more versatile.

Ironically, a further boost for WORM has been the success of CD-ROM. A WORM CD-ROM or CD-R recordable CD is a CD-sized 120 mm polycarbonate disc with a WORM film deposited on it. For CD-ROM sample discs CD-R recorder drives are now available from a number of manufacturers. The quality of these discs is adequate for samples and they are growing in popularity.

## 9.1 WORM technologies

Figure 9.1 summarises five technologies that are available for WORM media:

- hole forming;
- bubble forming;
- thermo-plastic (dye polymer);
- phase change;
- texture change ('moth eye').

All the technologies use a standard gallium arsenide laser to cause a permanent change to the reflectivity of the film. Laser powers of about 4 to 10 mW are required to record data, but only about 1 mW is needed to read it.

The hole-forming type uses a single film of tellurium, or tellurium-based alloy on a polycarbonate or glass substrate. Figure 9.2 shows the reflectivity of tellurium compared with the reflectivity at the hole where the tellurium has been 'boiled off'. The tellurium has a reflectivity of 35 %, whereas the glass surface in the hole has a reflectivity of only 4 % (Marchant, 1990).

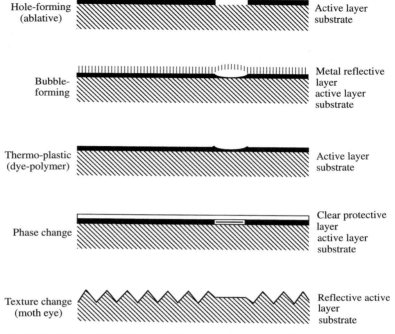

Hole-forming (ablative) — Active layer / substrate

Bubble-forming — Metal reflective layer / active layer / substrate

Thermo-plastic (dye-polymer) — Active layer / substrate

Phase change — Clear protective layer / active layer / substrate

Texture change (moth eye) — Reflective active layer / substrate

**Fig. 9.1** The different WORM optical film technologies to achieve 'write-once-read-many' properties.

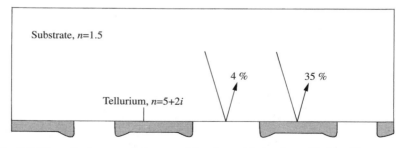

Substrate, $n=1.5$

Tellurium, $n=5+2i$

4 %    35 %

**Fig. 9.2** The reflectance variations resulting from ablation in a tellurium film.

Bubble forming is accomplished by heating the interface between a reflecting layer and an active plastic layer on a glass or plastic substrate. For example, titanium or gold films have been used on PMMA or polycarbonate.

Thermo-plastic or dye-polymers use dyes like carbocyanine or squarilium in polyvinyl acetate to absorb the infra-red and cause a surface distortion and the formation of a hollow-type 'defect' which can be used for data coding. This type is very popular for WORM media applications since it is very flexible.

Phase change media uses tellurium suboxide. This film is usually provided with a protective coating to prevent scratching. Matsushita, the Japanese company who developed this, uses a glassy tellurium oxide matrix in which very small tellurium particles are encapsulated. The reflectance at the substrate matrix interface is 18 %. After heating with the laser, the tellurium particles melt and coalesce, and recrystallise on cooling. The reflection from the recrystallised grains of tellurium is 26 %, so the recorded bits of data can be easily differentiated from the unrecorded ones. The reflectance variation of this phase change WORM media is shown in Figure 9.3.

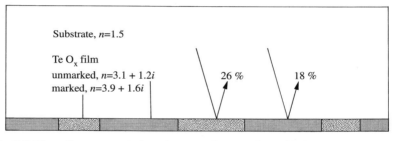

**Fig. 9.3** The reflectance variations due to coalescence in a tellurium sub-oxide film.

The texture change media resembles a moth's eye (ME), since the injection molded substrate has a unique microstructure which is generated during the mastering process by laser interferometry. When this ME substrate is coated with a platinum or gold film it has a reflectivity of 15 %. Upon heating the metal film-polycarbonate interface, the polycarbonate thermally degrades and the microstructure is relieved, so that a metal mirror with a higher reflectivity is formed as the recorded bit (see Fig. 9.1). Figure 9.4 shows the reflectivity change against the infra-red laser power: 4 mW of laser power is sufficient to produce the mirror bit and a reflectivity of over 50 % (Helfet and Pettigree, 1984).

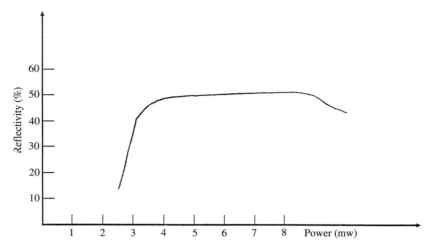

**Fig. 9.4** Reflectivity change on recording a bit on an ME (moth's eye) WORM disc rotating at 750 r.p.m.

Finally, one further technology should be mentioned and this is the alloying technique. In this case two or three layers are alloyed together. Examples are rhuthenium-silicon and antimony selenide-bismuth telluride alloys.

## 9.2  WORM optical heads

In this section, optical heads will only be mentioned briefly since WORM drives are very similar to the drives that are described elsewhere.

Figure 9.5 shows the WORM head. This reflecting sensing head can also be used for CD-ROM heads. It measures the intensity of the light reflected by the disc. The infra-red semiconductor diode source intensity is modulated electronically during writing to produce light pulses to code the data. After collimation by a lens the plane polarised light passes through a polarizing beam splitter and a quarter-wave plate, which makes it circularly polarised, before being focussed by the objective lens on to the disc. On reflection the light passes through the quarter wave plate again so that the polarised reflected light is orthoganol to the light originally transmitted by the beam splitter. The orthaganol light is then reflected by the beam splitter on to the detector system. Once again, as with the CD-ROM, a quadrant detector system is used for focussing the reflected light. However, because only a single incident beam is used for tracking, the same quadrant detector is used. When the beam is centred directly on a groove the first orders are identical and they interfere equally with the zeroth order.

**Fig. 9.5** A reflectivity-sensing WORM optical head.

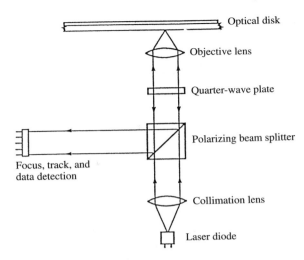

## 9.3 WORM tape

A UK company (ICI Image Data) has produced a flexible WORM media which it calls digital paper. It has a Melinex polyester substrate on which is coated a dye-polymer layer and a protective overcoat, as shown in Fig. 9.6 after Duffy and Gilson (1990). The dye absorbs infra-red light from the gallium arsenide laser and pits are formed, whose reflectivity is changed in comparison with the unwritten areas.

The first commercially available product based on this digital paper was an optical tape recorder designed by Creo. One twelve-inch reel of 35 mm-wide ICI digital paper tape can store 1000 GB of data code. The Creo 1003 optical

**Fig. 9.6** A schematic diagram of the structure of the ICI optical data storage 'digital paper'.

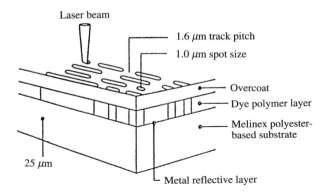

tape drive was designed specifically for mainframe applications such as archiving, medical imaging, seismic data logging and satellite image storage. The Creo drive has a SCSI interface and a data transfer rate of 3 MB per second. The bad news is that it takes an average of 28 seconds to access data stored on the tape. However, the good news is that it has an air-bearing scanner so that bits can be recorded across the tape. 32 data bits are recorded across the tape at the same time. Each sector of 80 KB has 16 KB for error correction, so errors are kept below 1 in $10^{12}$.

## 9.4 Present and future trends

CD-R is a WORM disc which complies with the CD standard format and is 120 mm in diameter. It is used whenever a small number of compact discs are required for test runs. It is so cheap to produce in comparison to the CD glass master disc that it is likely to be used in increasing volume. For banks and insurance companies, for example, only one or two copies of records are required, and here the WORM disc is an excellent storage medium because it has a long life.

Finally, the WORM card, or credit card, should be mentioned since this could replace the magnetic tape decrementing card for telephones and cash withdrawals from banks. The WORM card is very difficult to counterfeit, as expensive equipment must be used to manufacture it.

**Question**

Why is archival storage so important for WORM technology?

**Answer**

Archival storage requires only a small number of copies. It also needs a technology which is a permanent record which cannot be erased. CD-ROM provides a permanent record but is very expensive for small numbers of discs since mastering is relatively expensive. WORM discs can be produced at 10 % of the price of CD-ROM mastering. So up to ten copy discs can be produced for the same price.

CD-R is a WORM disc with the CD format and is gaining in popularity. A few copies of a **multimedia disc** containing audio, video, and data material, or a combination of all three, can be produced for long-term storage as a record or archive.

## References

Duffy, J and Gilson, R. (1990) unpublished conference presentation.
Helfet, P. R. and Pettigree, R. M. (1984) *TOC Conference* Plasmon Data Systems.
Marchant, A. B. (1990) *Optical Recording*, Addison-Wesley, 44–45.

# Magneto-optic media

**10**

## 10.1 Requirements

Magneto-optic media is attractive for use in optical recording systems because it can be erased and rewritten over one million times. For this reason it is used in a much wider range of applications than WORM media. However, unlike WORM, which can be read simply by measuring reflectivity changes, the signal received from a magneto-optic disc depends on both the reflectivity and the polar Kerr rotation (see Chapter 2).

In practice, the requirements for magneto-optic recording are even more complicated and include the following:

High figure of merit at the read laser wavelength;
Perpendicular magnetic anisotropy;
Square hysteresis loop;
High coercivity;
Curie temperature below 300°C so that it can be switched with a 30 mW laser;
Low thermal conductivity;
Low noise (see Chapter 12);
Write-read-erase cycle capability of at least one million times;
Stability against corrosion;
Amorphous or crystalline structure with nm grain size;
A simple low-cost media production process like sputtering.

The magnetic, optical and life properties will all be explained in more detail in the sections below.

## 10.2 Magneto-optic recording

Here we shall concentrate on the media and how to write, read and erase it. Later, in Chapter 11, the systems that are used in commercial drives will be described.

Figure 10.1 shows a cross-section of the magneto-optic media that is currently used in most drive systems. The laser light is focussed through the substrate on to the top surface of the magneto-optic (M-O) layer. Dielectric layers of aluminium nitride protect the M-O layer and the aluminium layer is used as a reflector. The magneto-optic signal is proportional to the figure of merit $R\theta_K^2$ and so a high reflectivity is obtained by using a quarter wavelength for the aluminium nitride layer next to the substrate.

**Fig. 10.1** Cross-section of commercially-available magneto-optic media, showing four thin film layers on a glass or plastic substrate.

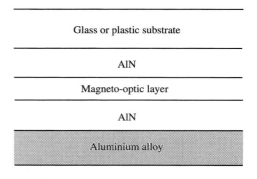

Figure 10.2 compares $R^{1/2}\theta_k$ as a function of wavelength for the commercial rare earth transition metal media terbium-iron-cobalt amorphous alloy

**Fig. 10.2** Variation of the figure of merit (Kerr angle) of commercially-available terbium–iron–cobalt magneto-optic media against wavelength. For comparison, platinum–cobalt multilayers and garnet are also shown.

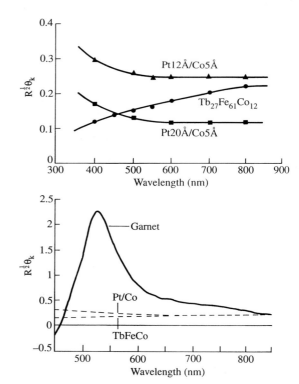

(Takahashi, *et al.*, 1991) with crystalline platinum-cobalt multilayers and garnet (Fujitsu, private correspondence and Shono, *et al.*, 1991). If the figure of merit were more important then platinum-cobalt and especially garnet would be far superior to the rare earth transition metals, particularly in the blue and the green part of the spectrum. Here the density is higher, since the wavelength is directly proportional to the focussed spot diameter.

To protect the M-O writable discs and layers, two are glued together in the manner shown in Fig. 10.3. The M-O digital data is written on the land area between the grooves which are 1.6 μm apart. The grooves are used for tracking in a similar way to the compact disc. As with the compact disc pits, the grooves are injection molded into polycarbonate substrates.

**Fig. 10.3** The construction of a writable magneto-optic commercial disc in which two discs are bonded together so that the recording layers are protected by the polycarbonate (PC) substrates. Enlarged view shows V- or U-shaped grooves/pits.

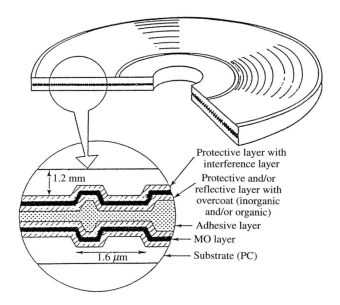

Protective layer with interference layer

Protective and/or reflective layer with overcoat (inorganic and/or organic)

Adhesive layer

MO layer

Substrate (PC)

1.2 mm

1.6 μm

**Disc sectoring** can be achieved in four ways by moulding tracks in the manner shown in Fig. 10.4. The upper half-disc shows disc sectoring for constant angular velocity (CAV) or modified CAV (MCAV) where a variable data rate is used. The lower half-disc illustrates the higher density that can be achieved with constant linear velocity CLV (continuous servo) and modified (MCLV) with banded servo.

Finally, to complete this M-O recording section, the recording, reading and erasing process will be simply described before the magnetic and optical properties are discussed in detail.

**Fig. 10.4** Disc sectoring methods for constant
angular velocity and constant linear velocity.

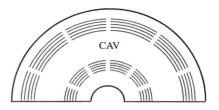

CAV   -fixed disk speed, fixed data rate
MCAV -fixed disk speed, variable data rate

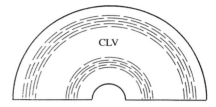

CLV   -variable disk speed (continuous servo)
MCLV -variable disk speed (banded servo)

Figure 10.5 shows the first step in the recording process. The laser beam
heats up a spot about one micron wide until the temperature is just below the
Curie temperature (see below). The magnetic domain field strength is reduced
to close to zero at this point. An external applied magnetic field, using an
electromagnet or a permanent magnet, is then used to reverse the domains to
record a digital one or a zero, which depends on the data coding used. A
semiconductor laser power of 30 mW is required for writing since after

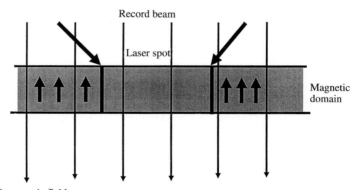

**Fig. 10.5** Recording information on a magneto-optic disc showing reversal of the
magnetic domain by an applied magnetic field during localized heating close to the
Curie temperature.

passing through the optical head the power on the disc is reduced to about 10 mW. The recording field has to be in the region of 300 oersteds.

To read the recorded data the Kerr effect is used as Fig. 10.6 illustrates. The *E* vector rotation for the elliptically polarized light is clockwise for the newly recorded bit, say a 'one', and anticlockwise for the 'zero', recorded vertically through the film. The rotation occurs on reflection and this Kerr effect will be discussed in more detail below. The Kerr effect is observed regardless of whether the applied magnetic field is present or not. Consequently, it is possible to use either a small permanent magnet, which is rotatable to reverse the field, or an electromagnet in which the coil current can be reversed.

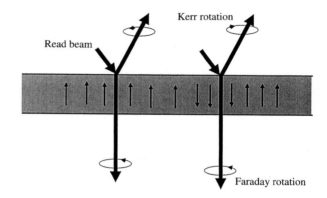

**Fig. 10.6** Using either the polar Kerr or the Faraday rotation to read information from a magneto-optic disc.

Figure 10.7 shows how the data can be erased by heating just below the Curie temperature, so that with the applied field reversed the domains can be switched and returned to their original vertical polarization.

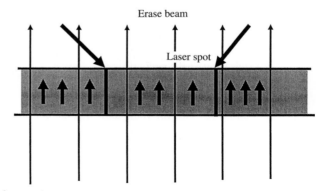

**Fig. 10.7** Erasing information on a magneto-optic disc by field reversal during localized heating.

## 10.3 Magnetic properties as a function of temperature

Choudari, *et al.* (1976) discovered the amorphous rare earth transition metal alloys in 1976 and these are used commercially at the time of writing. Terbium-iron-cobalt and gadolinium-terbium-iron are both available. Amorphous films are more elastic than crystalline films and the composition can be changed to alter the temperature performance and resulting benefits to magnetic recording.

Rare earth transition metal alloys have magnetic properties that change rapidly with temperature according to Marchant (1990). They are ferrimagnetic and this means that the rare earth (RE) atoms are magnetised in the opposition direction from the transition metal (TM) atoms. These opposite magnetizations are shown as a function of temperature in Fig. 10.8. They both reduce as the temperature increases until they reach zero at the Curie temperature.

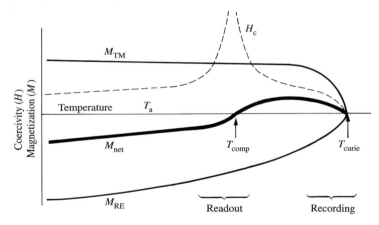

**Fig. 10.8** Qualitative magnetic/temperature characteristics of ferrimagnetic rare earth (RE) and transition metal (TM) materials. Bold line shows net magnetization; broken line shows net coercivity.

In most cases the $M_{TM}$ and $M_{RE}$ cancel out at $T_{comp}$, the compensation point. In theory, compensation point recording can be used as well as Curie point recording, but in practice Curie point recording is less 'noisy' because the control is more accurate. The concentration of the RE/TM alloy is critical since this will affect the noise, especially if $T_{comp}$ is close to $T_{curie}$. Reasonable noise values can be obtained if $T_{comp}$ is set in the centre between ambient temperature, $T_a$, and the Curie point temperature.

If only two elements are present then control of the compensation point is quite difficult. This is shown by Fig. 10.9, which compares terbium-dominated terbium-iron alloys with iron-dominated ones. A change from 19 % atomic terbium to 23 % atomic terbium alters the compensation temperature from about $-100°C$ to $+80°C$. For this reason a third element, cobalt or gadolinium, is added to the alloy to provide better control and a lowering of the Curie temperature to about 200°C.

**Fig. 10.9** Three representative magnetization curves for terbium–iron alloys showing that the compensation point depends strongly on the composition.

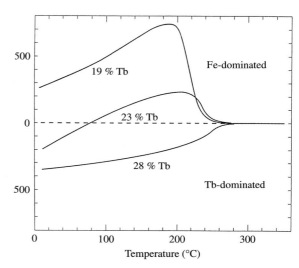

### Question

Why is an amorphous ferrimagnetic alloy like terbium-iron-cobalt useful for magneto-optic recording?

### Answer

The ideal magneto-optic recording material needs a Curie temperature of 150 to 200°C. By careful choice of composition the terbium reduces the Curie temperature from the high values of iron and cobalt until the ideal value is achieved at a composition in the region of 28 % terbium 66 % iron 6 % cobalt. These alloys are amorphous and have vertical magnetic anisotropy. The amorphous nature is important, because a wide range of substrates can be used and readout and recording noise caused by crystallite boundaries does not occur. There is some degree of short-range order of the atoms since the magnetisation of the rare earth atoms compensates for the opposite magnetisation of the transition metals. Hence, they are ferrimagnetic (see Fig. 10.8).

## 10.4 Substrate variations

The rare earth transition metal materials all have an amorphous structure and so they can be deposited on a very wide range of substrates. Figure 10.10 summarizes the coercivities measured using the polar Kerr magnetometer on eighteen different substrates, ranging from polycarbonate of the same optical quality that is used for commercial M-O discs all the way through a variety of metals to glass and single crystal silicon. In the vast majority very high coercivities are obtained, so this means that amorphous M-O material has the potential to be used in a wide range of applications.

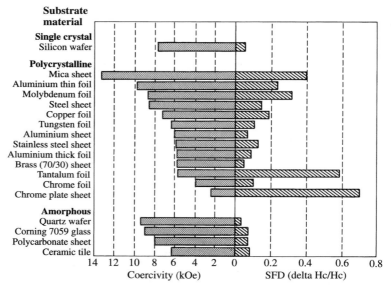

**Fig. 10.10** Dependence of coercivity and switching field distribution (SFD) for eighteen different substances as measured with terbium–iron–cobalt alloy. (Courtesy: Keele University.)

## 10.5 Optical and magnetic modulation compared

The current commercial rare earth transition metal media suffer from overwrite problems unless each track is erased twice. This double erasure is time consuming when high speed is required.

To overcome this problem, optical or magnetic modulation can be used. Figure 10.11 shows the disc structure for **magnetic field modulation** (MFM) overwrite. The magnetic head that is used to create the magnetic field for recording and erasing domains flies very close to the thin film surface, so an abrasion-resistant coat is deposited on top of the vapour barrier coating. The abrasion resistant coating contains alumina, a surface-active agent and a lubricant in a resin that can be UV-cured in one or two seconds.

**Fig. 10.11** Disc structure for magnetic field modulation (MFM) overwrite media.

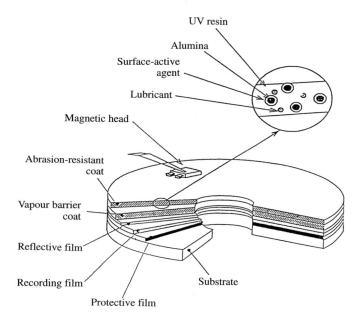

The Sony recordable audio Mini-disc uses this single-sided MFM technology and has a polycarbonate substrate. Figure 10.12 illustrates the Sony Mini-disc in comparison with standard CD-ROM and WORM discs.

**Laser intensity modulation** (LIM) overwrite uses generally at least one additional magnetic layer. The simplest form of LIM was invented by Nikon and is illustrated in Figure 10.13 which shows the two-part magneto-optic layer for writing and reading. To write, the highest laser power raises both films above $T_{curie}$. Both layers can then have their domains orientated by magnet 1. Magnet 2 re-orientates the reference layer. To erase the memory layer, heat to temperatures between $T_{curie}$ for the reference layer $T_{cr}$, and $T_{curie}$ for the memory layer ($T_{cm}$). The memory layer will then reorientate to the reference layer domain alignment by exchange interaction. Overwriting is achieved by pulsing the laser to a high power during the erase cycle. Finally, as with conventional M-O recording, a low power of 1 to 3 mW is required at the disc to read the data.

**Fig. 10.12** Selection of data discs: standard CD-ROM, top left; Sony Mini-disc (MFM media), top right; WORM disc, bottom right; stamper disc, bottom left.

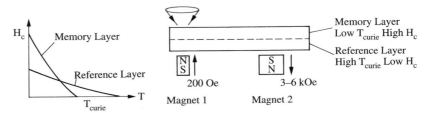

**Fig. 10.13** The mechanism for laser intensity modulation (LIM) overwrite media.

Figure 10.13 also shows the relative values of $T_{curie}$ and coercivity ($H_c$) for the reference and memory layers.

A comparison of the recorded marks or magneto-optic domains for MFM and LIM recording is shown in Fig. 10.14. LIM has a variable domain width whereas MFM has crescent-shaped domains and hence more inter-symbol interference.

The creation of the two different types of domain is clearly illustrated in Fig. 10.15 (Watanabe, 1989). In MFM the light power stays at the high level during recording and the magnetic field is pulsed. This means that the film is already heated up at $t_0$ and does not stop forming the domain until the field is

**Fig. 10.14** Comparison of the recorded domains for MFM and LIM recording.

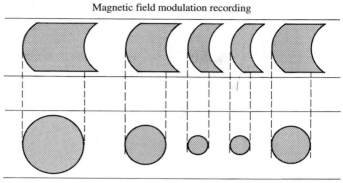

Magnetic field modulation recording

Light intensity modulation recording

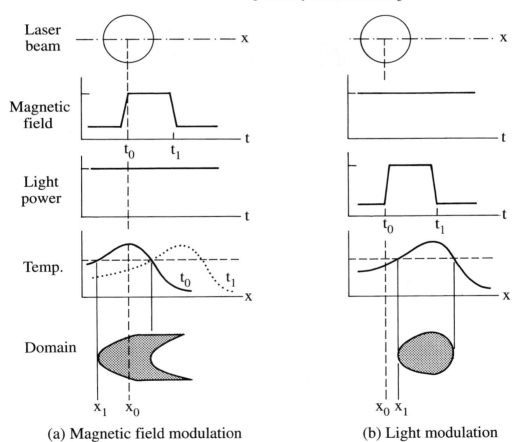

(a) Magnetic field modulation

(b) Light modulation

**Fig. 10.15** Model of the recording process for (a) MFM and (b) LIM media.

switched off. For LIM, however, the light power is pulsed, so the laser spot takes a little time to heat up the film, and hence the domain is circular and not as long nor as wide as for a small bit size. Edge shift of the written domain, between $X_1$ and $X_0$ is caused by the temperature distribution on the disc, depending on recording conditions such as the write laser power, linear velocity, laser beam spot diameter and shape and the thermal conductivity.

In summary, MFM and LIM are compared in Table 10.1, according to C. Hall. MFM can only be single sided so an abrasion resistant layer is required. Because MFM uses a low flying magnetic head, head crashes are possible. With a magnetic field high frequency modulation is difficult. The fixed domain width with MFM does make it easier to handle the output signals reflected from the disc. LIM, on the other hand, can use a double sided disc formed as in Figure 10.3. The lens is 1 mm from the substrate, so head crashes with the substrate are unlikely. The variable domain width means a large variation in signals, but intersymbol interference is absent in LIM.

**Table 10.1**
Comparison — MFM and LIM overwrite

| MFM | LIM |
| --- | --- |
| Single-sided disc | Double-sided disc |
| Abrasion-resistant layer required | Exchange-coupled layer structure |
| Potential head crash | Potential thermal degradation |
| Difficult to modulate at high frequency | Easy to modulate at high frequency |
| Fixed domain width | Variable domain width |
| Large recording power margin | Small recording power margin |
| Already commercialized | |

# 10.6 Media life studies

Media degradation can be caused by oxidation and structural changes at 300°C in commercial terbium-iron-cobalt alloys. Oxidation can be avoided by using pin-hole-free silicon or aluminium nitride protective layers in a three-layer structure. High temperatures of 300°C can be avoided by using a thermo-diffusing (TD) layer of aluminium alloy or pure aluminium. The temperature profile is shown for three-layer and the four-layer discs in Fig. 10.16 after Ogihara et al. (1989). Figure 10.16(a) shows that a temperature of over 300°C occurs with the tri-layered disc in order to generate a 1.2 $\mu$m track width of temperatures greater than 180°C. At this temperature the domains could be switched with an applied field of 400 Oersteds. In

Fig. 10.16(b) the same track width and temperature greater than 180 degrees was achieved, but with a maximum temperature of 260 degrees when the TD layer was added. This confirms the heat diffusion model predicted in Fig. 10.17.

**Fig. 10.16** Temperature distribution during the erase process on the outermost track at a disc speed of 15 m/s (a) tri-layered (b) quadri-layered discs. Lines of equal temperature are marked in 20°C intervals. The moving region in excess of 180°C is shown shaded.

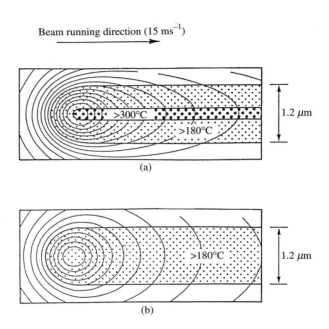

**Fig. 10.17** Schematic illustration of the diffusion of heat during erase or write processes with (a) tri-layered and (b) quadri-layered discs, showing the improved heat flow with the added thermo-diffusing layer TD.

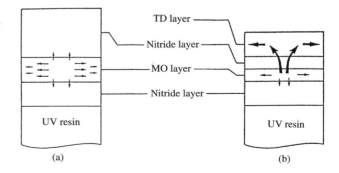

For the four- and three-layer discs shown in Fig. 10.17, no degradation in carrier signal level was seen after more than $10^7$ write-read-erase cycles, as Fig. 10.18 shows in tests reported by Ogihara. The tri-layered disc, on the other hand, began to degrade after $10^3$ cycles.

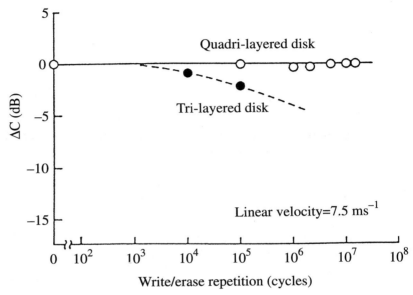

**Fig. 10.18** Comparison of the life performance of the quadri-layer and tri-layer discs.

Five-layer structures, which include an organic protective layer on top of the aluminium alloy show in Figure 10.1, with corrosion inhibitors added to the rare-earth transition-metal alloys, have further improved life expectancy.

New methods of accelerated life testing are under development. One of these is the 24 hour Z/AD cycle. As Fig. 10.19 shows, the temperature is cycled from 20 to 70°C twice and then down to –20°C. Over 85 % relative humidity is used, except when the temperature goes negative and condensation is avoided by not controlling the humidity at the high level. Z/AD cycling speeds up the ingress of moisture, and hence oxygen, to cause rapid degradation. It also stimulates what can happen during transport of the media around the world in aircraft or at sea aboard ships.

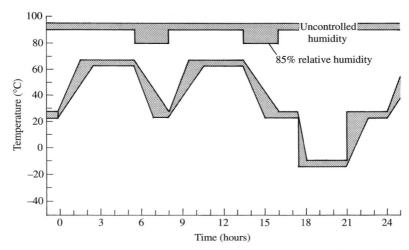

**Fig. 10.19** Temperature and relative humidity limits for the Z/AD cycle used for life stressing magneto-optic media.

# 10.7 Future trends

Cobalt-platinum alloy or multi-layer structures (MLS) are most likely to replace rare earth transition-metals because they are chemically more stable and have a higher figure of merit at visible laser wavelengths. They do suffer from two disadvantages: high Curie temperature and crystalline structure.

Further work is required to reduce the Curie temperature for MLS since for visible lasers (except for red) high powers are not yet available, and there is evidence of inter-diffusion between layers occurring at 300°C.

Garnet films are still some way off because they cannot be produced on polycarbonate substrates, and they produce grain noise owing to the crystallites. Nevertheless, they still remain the most promising alternative to platinum-cobalt MLS.

# References

Choudri, P., Coumo, J., Gambino, R., and McGuire, T. (1976) US Patent 3949387.

Marchant, A. B. (1990) *Optical Recording*, Addison-Wesley, p. 76.

Murray, W. P. (1991) *Journal of the Magnetics Society of Japan*, **15**, Suppl. S1, pp. 357–360.

Ogihara, N., *et al.* (1989) *Japanese Journal of Applied Physics*, **28**, Suppl. 28–3, pp. 61–6.

Shono, K., Kuroda, S., and Ogawa, S. (1991) *Journal of the Magnetics Society of Japan*, **15**, Suppl. S1, p. 290.

Takahashi, M., Nakamura, J., Nijhara, T., and Tatsuno, K. (1991) Magneto-optic recording information systems (MORIS), *Journal of the Magnetics Society of Japan*, **15**, Suppl. S1, p. 49.

Watanbe, H., Nakao, T., Maeda, T., Saito, A., and Nagai, S. (1989) Optical data storage topical meeting, *Society of Photo-optical Instrumentation Engineers*, **1078**, p. 296.

# Magneto-optical drive systems

**11**

The magneto-optic drive system came on the market in 1989 and the first European standards were published in 1991. The first Sony drive was extremely reliable, so this will be featured to illustrate how the drives operate.

A more recent development by Maxoptix will then be featured, since this has led to a very fast access time of 35 ms. This uses a lightweight optical head, which will be described in some detail.

All magneto-optic drives suffer from noise from the laser, the media and the electronics, and these additive noise sources will be summarized in a later chapter.

Eight commercial 130 mm drives will be compared in terms of capacity, data transfer rate, average access time and disc rotation speed.

## 11.1 The Sony SMO–D501 disc drive

The outline specification for the Sony 130 mm disc drive is shown in Table 11.1. The total data capacity is 650 Mbytes for both sides of the disc. 512- or

**Table 11.1**
**The Sony (130 mm) disc drive specification SMD–D501**

| Item | Specification | Remarks |
|------|---------------|---------|
| Mechanical dimensions (mm) | 82.5(H) × 146.0(W) × 203.2(D) | Not including the connector |
| Weight | 3 Kg | |
| DC Voltage | + 5 V ± 5 % | Ripple voltage <50 m Vpp |
| | + 12 V ± 5 % | Ripple voltage <100 m Vpp |
| DC Current at +5 V | 1 A (max.) | |
| DC Current at +12 V | 1 A (typ.) 2.5 A (max.) | at spinning up and seeking |
| Operating Mount | Horizontal or Vertical | |
| Laser Diode type | Semiconductor Laser | Class 1 (IEC. 825) GaA1As |
| Wave Length | 790 nm continuous | |
| Output Power | 30 mW max | |
| Beam Divergence | 60° ± 1.5° | |

**Table 11.1   (*cont.*):**

| | | |
|---|---|---|
| **Environmental:** | | |
| Operating temperature | 5–40°C | Recommend to use a cooling fan |
| Relative humidity | 10–80% | Not condensing |
| Max. wet-bulb temp. | 29°C | |
| Temperature gradient | 10°C/Hour | |
| Storage temperature | −30°C–60°C | |
| Storage RH | 5–90 % | |
| Vibration, operating | 0.25 g | 5 to 500 Hz sine sweep |
| Non-Operating | 1 g | 5 to 500 Hz sine sweep |
| Shock, operating | 40 g | 3 ms half sine pulse |
| Non-operating | 89 g | 3 ms half sine pulse |
| | 30 g | 23 ms trapezoidal pulse |
| Drive interface | Modified ESDI | ESDI: enhanced small device interface |
| Disc type | 130 mm (5.25 inch) double-sided MO disk with cartridge | Formatted Address: Track 0 Sector 0 to Track 18750 Sector 16 or 30 |
| Disc capacity: | | |
| Formatted (per side) | 325 Mbytes | (1024 bytes/sector) |
| | 294 Mbytes | (512 bytes/sector) |
| Unformatted (per side) | 433.5 Mbytes | (1024 bytes/sector) |
| | 433.6 Mbytes | (512 bytes/sector) |
| Disc format | Continuous/composite | ISO standard type A |
| Rotational mode | CAV | CAV: constant angular velocity |
| Write/read | 2 | Double-sided |
| Bytes per sector | 1024 or 512 | |
| Sectors per track | 17 or 31 | |
| Total tracks per side | 18751 | |
| Rotational speed | 2400 rpm | CAV |
| Average latency | 12.5 ms | |
| Seek Time: | | |
| Short stroke (+64 Tracks) | 22 msec | including overhead time |
| Average | 95 msec | |
| Full Stroke | 184 msec | |
| Data transfer rate | 7.40 Mbps | bps = bits per second |
| User data transfer rate | 680 Kbytes/sec | (1024 bytes/sector) |
| | 620 Kbytes/sec | (512 bytes/sector) |
| Loading time | 6.5 sec (Typ.) | Including spin-up time |
| Unloading time | 3.5 sec (Typ.) | Including spin-down time |
| Bias magnet rotation time | 20 msec (Typ.) | Without SMO - C501 Defined in ESDI level (drive ready situation) |

1024-byte sectors can be used. 433 Mbytes of user data is available for storing erasable information on the disc.

The rotational speed is 2400 rpm and since it has a constant angular velocity (CAV) the speed is always kept constant. The average time taken to access data on the disc is 95 ms and this is limited by the weight and the mechanical movement of the actuator that holds the objective lens which finally focuses the laser light on to the magneto-optic disc.

The user data transfer rate is 680 Kbytes s$^{-1}$ for a 1024-byte sector.

The disc is automatically loaded in 6.5 s and this includes the time taken for the motor to reach 2400 rpm.

A permanent bias magnet is used to magnetise the magneto-optic film that is heated to the Curie temperature under the laser light spot. This magnet takes 20 ms to turn over and reverse the field.

An infra-red aluminium gallium arsenide laser is used with a single mode emission wavelength of 790 nm and a maximum power of 30 mW. In practice, the write and the erase power used will be about 8 mW only and the read power of about 2 mW will be the normal continuous operating power.

The block diagram of the system is shown in Fig. 11.1. The laser diode and the photodetectors are both in the optical head. The laser diode (LD) driver and the RF and servo-preamplifiers are all built into this head. The analogue servo systems drive the spindle motor that rotates the disc; the slide motor then given course movement to the optical head, the tracking location and the fine focus.

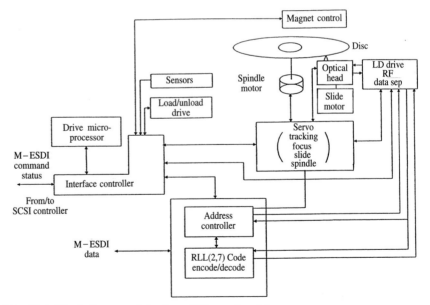

**Fig. 11.1** Block diagram of the Sony 130 mm disc drive.

Digital run length limited coding, RLL (2, 7), is used, and more will be said of this shortly. Cross-interleaved Reed-Solomon (CIRC) error-correction coding is added, since this was very successfully developed for the compact disc.

Finally, the interface controller is linked to the SCSI (small computer systems interface) controller, and this makes the drive compatible with a very wide range of computer systems.

## 11.2 Coding and the RLL code

Figure 11.2 indicates the coding steps that are required in an optical recording system. After the binary input data has had the error correction code added, it is converted into an RLL modulation code.

**Fig. 11.2** Coding steps required in an optical recording system.

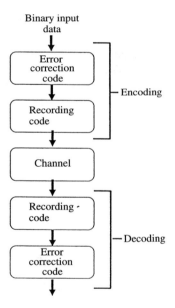

The RLL code offers better dc. stability, accurate timing synchronisation and can support a higher data density. It is defined by code rules that convert the data into a longer sequence of channel bits during encoding. RLL (2, 7) code has at least 2 '0's following each channel bit '1' and at most 7 '0's. In addition, the constant channel clock must correspond to a constant data rate.

Table 11.2 shows how the input data bits relate to the RLL (2, 7) channel-bits. Pit position recording is used by the Sony drive, so that all the recorded regions are identical in size and the spacing between them represents the distance between the '1's.

Table 11.2
The RLL (2, 7) modulation code

| DATA BITS | CHANNEL-BIT (2, 7) |
|---|---|
| 10 | 0100 |
| 11 | 1000 |
| 000 | 000100 |
| 010 | 100100 |
| 011 | 001000 |
| 0010 | 00100100 |
| 0011 | 000010000 |

Figure 11.3 illustrates pit position recording for a RLL (2, 7) channel-bit sequence.

**Fig. 11.3** Pit position recording for a RLL(2, 7) channel-bit sequence.

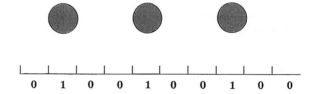

0 1 0 0 1 0 0 1 0 0

# 11.3 The Maxoptix Drive

A major breakthrough was achieved by Maxoptrix by using a lightweight fine focus actuator separated from a coarse actuator and the rest of the fixed optics (Johann and Burroughs, 1991). The fine focus actuator weighs only 1.8 grams and is shown in Fig. 11.4. The objective biconvex lens and a mirror are the only optical components. The objective lens is a normal two-dimensional type with two sets of coils and permanent magnets. The position of this actuator is controlled by the tracking error signal from the detectors.

The coarse actuator keeps the parallel beam centred on the objective lens with a position sensor pair that measures the light emitting diode (LED) light reflected from the mirror mounted on the fine focus actuator. The coarse actuator weighs approximately 60 g.

The actuator incident beam is diffracted at the groove edge as Fig. 11.5 shows. The groove is in the form of a continuous spiral with a 1.6 μm track-to-track spacing. The zeroth order reflected beam and the two first order diffracted beams are focussed on to the PIN silicon detector array. The detector consists of six PIN detectors on a single slice of silicon.

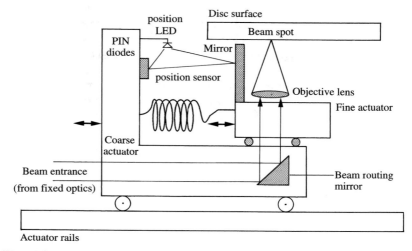

**Fig. 11.4** Coarse and fine actuator of the Maxoptix drive.

**Fig. 11.5** Tracking error signal
generator, showing servo sensor (top).

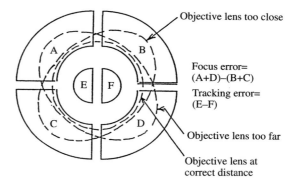

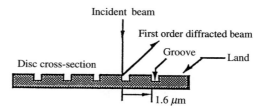

The focussing is carried out by the astigmatic method using the detectors A, B, C and D. The focus error (*FE*) signal is defined by: $FE = (A+B) - (B+C)$. To achieve $FE = 0$ and hence perfect focus a threshold r.f. signal is obtained by summing $A + B + C + D$ and this is used to activate the fine focus servo.

The first order diffracted light beams strike the detectors E and F so that one light beam increases in intensity as the other decreases. Hence the term push-pull tracking. The tracking error (*TE*) is given by: $TE = E - F$.

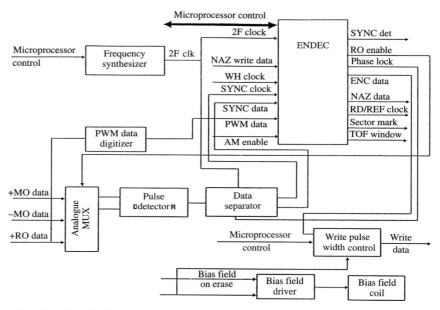

**Fig. 11.6** Read/write system in the Maxoptix drive.

Now, moving on to the read-write system, this is described by the block diagram in Fig. 11.6. The read-write channel operates over data rates varying from 6.8 Mbs$^{-1}$ at inside diameter to 13.1 Mbs$^{-1}$ at the outside diameter. Zoned recording increases the capacity of the disc from 650 Mb to 1 Gb. Both formats can be used in the drive. The externally applied field used to write and erase is an electromagnet as opposed to a permanent magnet that Sony used. The pulse detector is shown in Fig. 11.7. This combines peak detection with level qualification. The input to the AGC amplifier is multiplexed between the signals from the preformatted data preamplifiers and the magneto-optic data preamplifiers, depending on whether headers or data are being read.

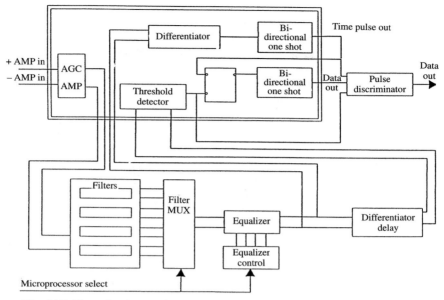

**Fig. 11.7** The pulse detector in the Maxoptix drive.

Resolution at around 50 % in the optical recording channel is much worse than in most magnetic recording channels. Equalisation is performed using a cosine equaliser $(1 - K\cos\omega t)$ with $K = 0.8$. It consists of a delay line with selectable taps and tap weights fed into a summing amplifier. The time constant of the equaliser varies from 100 ns at the inside diameter to 50 ns at the outside. Figure 11.8 shows an equalisation curve. A compromise was

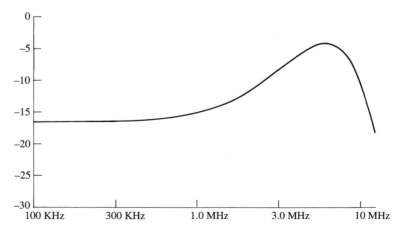

**Fig. 11.8** An equalizer curve: voltage output in dB against frequency.

made between the desire for 100 % equalisation and the need to control the amplitude of the side lobes introduced by the equalisation process. After equalisation a 50 % input resolution is transferred into a nominal 90 % resolution at the input to the gate channel.

Figure 11.9 shows that the recovered signal from the magneto-optic disc has significant d.c. content. To solve this problem the signal was differentiated before level qualification and AGC detection.

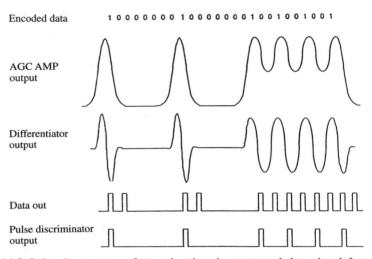

**Fig. 11.9** Pulse detector waveforms showing the recovered data signal from the magneto-optic disc.

# 11.4  Current commercial 130 mm magneto-optic drives

Table 11.3 gives a summary of eight of the magneto-optic 130 mm drives available in 1993. The latest development is the double density disc with a two-sided capacity of 1.3 GB that has been announced by IBM. Sony and Maxoptic have also just announced high capacity discs but full details were not available at the time of going to press. This increase in density with the banded-format technique has simply been achieved by increasing the linear recording density while decreasing track spacing.

The high speed drives made by Hewlett Packard and Ricoh have faster average access times. Speeds higher than this of up to 5400 rpm have been reported, so this will mean an even faster access time.

**Table 11.3**
**1993 Commercial 130 mm magneto-optic drives**
(*Data produced by Williams Associates*)

| Supplier | Model | Capacity per side (MB) | Mode | Data rate (KBS$^{-1}$) | Av. Access (ms) | Speed (rpm) |
|---|---|---|---|---|---|---|
| IBM | 063 – C2 | 650 | CAV | 1561 | 67.5 | 2400 |
| | 0632 | 325 | CAV | 696 | 82.5 | 2400 |
| Hewlett-Packard | C1716C/F | 325 | CAV | 1000 | 35.3 | 3600 |
| Maxoptix | TMT-11 | 322 | CAV | 850 | 48.6 | 2200 |
| | | 500 | MCAV | 1250 | | |
| Ricoh | RS – 9200EX | 326 | CAV | 1000 | 36.3 | 3600 |
| Sharp | JY – 750 | 326 | CAV | 870 | 50.0 | 3000 |
| Sony | SMO – E511 | 325 | CAV | 680 | 82.5 | 2400 |

The 1 GB Maxoptic disc normally has a glass substrate and there is little doubt that for some applications glass substrates are highly desirable.

Further increases in density of the drive will depend on new technology and some possibilities are outlined in the final chapter.

# Reference

Johann, D., and Burroughs, A. (1991) *IEEE* Transactions on Magnetics, **27**, p. 4496.

## Question

What is the difference between the average access time and the average seek time for an optical head?

## Answer

The average **latency** is the difference. For example, the new IBM magneto-optic drive 063–C2 has an average seek time of 55 ms, an average latency of 12.5 ms and therefore an average access time of 67.5 ms (see Table 11.3). The seek time is the average time the head takes to move to the required track. Latency is the time taken to fine focus and settle in the correct position on the track. Access time is the sum of these two.

# Noise in magneto-optic drive systems

# 12

Noise sources are additive and the magneto-optic drive systems have more noise sources than the CD-ROM or the WORM media. The CD-ROM, with only one reflective layer, is the simplest and has the lowest noise. The WORM system varies, depending on which of the four or five different technologies is used. In one case, the moth's eye, or the mirror bit recording technique, manufactured by Plasmon Data Systems, the noise is almost as low as for the CD-ROM, but for others it is higher.

For magneto-optic recording systems we shall consider three main sources of noise:

- Electronic
- Laser
- Media

After defining how carrier-to-noise levels are measured, each of these noise sources will be discussed in turn.

## 12.1 Carrier-to-noise definition

After writing 1.3 $\mu$m bits of information on a track with a laser power of about 10 mW, a frequency of 3.7 MHz, a 200 Oersted field, a disc rotation speed of 1800 rpm and a track-to-track spacing of 1.6 $\mu$m, the narrow-band signal-to-noise ratio, called carrier-to-noise ratio, can be measured. The data is read with a laser power of 1.5 mW, a spectrum analyser with a centre frequency of 3.7 MHz (carrier) and a bandwidth of 30 kHz. The amplitudes of the signal-to-noise figures are measured at $f_c$ as Figure 12.1 shows.

The narrow-band signal-to-noise ratio is expressed as $20\log \times$ signal level/noise level. This ratio is required by the ECMA–153 standard to be greater than 45 dB for all the tracks in the 'User Data Zone' and for all phase differences between $-15°$ and $+15°$ in the standard optical system.

## 12.2 Electronic noise

There are six types of electronic noise:

- Shot noise;
- Johnson (thermal) noise;

**Fig. 12.1** Carrier-to-noise ratio at the centre frequency, $f_c$, of 3.7 MHz.

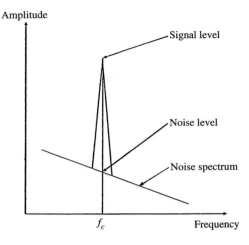

- preamplifier noise;
- dark current noise;
- amplifier bias current noise;
- $1/f$ noise.

The first two can be quantified. They are both white noise sources and are therefore constant with frequency. Shot noise is due to the quantum and probalistic nature of the photons impinging on the p-i-n detector.

The noise current, $I_{ns}$, is given by:

(12.2a)
$$I_{ns} = \sqrt{2eIB}$$

where $I$ is the current through the differential amplifier and $B$ is the bandwidth of the system and $e$ is the single electron charge.

The signal output from the photodiodes is given by:

(12.2b)
$$I = \eta P_o R\sin^2\theta_k$$

where $\eta$ is the responsitivity of the detector, $Po$ is the average laser power incident on the disc, $R$ is the reflectivity of the disc and $\theta k$ is the Kerr rotation angle. For magneto-optic recording the Kerr rotation angle is only about 0.3°. Multilayer optical structures have been used to increase this angle but they reduce the reflectivity. If the reflectance is low during reading and the absorption is high, there is a danger of the laser heating up the magneto-optic layer and causing corruption of the data.

The Shot noise gain of an avalanche photodiode is higher than the signal gain, so a p-i-n detector is preferred for magneto-optical drive systems.

In summary of the Shot noise, from the equations above, the signal-to-noise ratio, $SNR$, is given by:

(12.2c)
$$SNR_{SHOT} = 10\log_{10}(2P_o R\sin^2\theta_k)/eB$$

Johnson noise occurs in electronic devices and is due to the thermal fluctuations of electrons. It occurs in the load resistor of the optical detector. Johnson noise current, $I_J$, is:

(12.2d)
$$I_J = \sqrt{\frac{4kTB}{R_L}}$$

where $k$ is Boltzmann's constant, $T$ is the temperature and $R_L$ is the load resistor restance. The Johnson noise power is given by:

(12.2e)
$$N_J = \frac{1}{\eta}\sqrt{\frac{4kTB}{R_L}}$$

# 12.3 Semiconductor laser noise

There are four noise sources in this category:

- spontaneous light emission;
- variations in the laser input signal current;
- temperature variations;
- laser feedback.

Spontaneous emission, in which electrons fall from the conduction band to the valence band of gallium arsenide or the semiconductor alloy, happens randomly but is only a very small noise source at laser power levels above the threshold for lasing action.

Variations in the input signal current obviously depend on the quality of the power supply that is used. It is a simple matter to keep this noise source to a very low level.

Temperature variations can be kept to a minimum by mounting the laser on a heat sink. The laser diode power/current characteristic is very critically dependant on temperature, so it is essential that any operating optical drive is not placed in an environment where the temperature will fluctuate by large amounts of say five to ten degrees centigrade.

The most critical noise source is laser feedback noise and this will now be discussed in detail. At high injection currents above the laser threshold current the laser displays single mode behaviour and emits only one very short laser line in which all the light waves emitted are in phase or coherent with each other. However, the laser can become multi-mode and emit lines at different wavelengths if light is fed back into the laser semiconductor chip.

The laser can 'hop' wavelengths, because extra interference emission lines are created by this additional light coming into the laser. The interference occurs in the optical system because the optical disc acts as a reflector and, together with the semiconductor chip emitting mirror facet, this generates an external laser cavity. The wavelengths due to the external cavity are given by:

(12.3)
$$\Delta\lambda_1 = \frac{\lambda^2}{2L_{ext}}$$

since the refraction index of air is unity.

Figure 12.2 shows the wavelengths due to the external and the internal cavity. If the external path is fixed, so that the actual internal wavelength at which the laser is lasing is midway between the two external modes, then the laser can emit at either of these modes and the wavelength varies randomly between them and so produces noise.

**Fig. 12.2** Internal and external modes due to laser feedback noise.

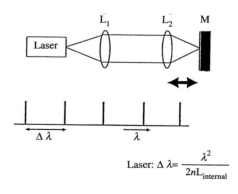

Laser: $\Delta\lambda = \dfrac{\lambda^2}{2nL_{internal}}$

External: $\Delta\lambda_1 = \dfrac{\lambda^2}{2L_{external}}$

Figure 12.3 shows the laser feedback noise against C, a parameter which combines the value of the feedback and the closeness of the internal and external cavity modes. There are three noise peaks that occur in this figure. The first small peak is at C = 1, where the optical feedback is about 0.01 %. This occurs because the laser hops between more than two modes. The final large peak occurs when the laser hops between internal modes in addition to external modes. The laser noise is constant just below the first peak, and this is an excellent operating point.

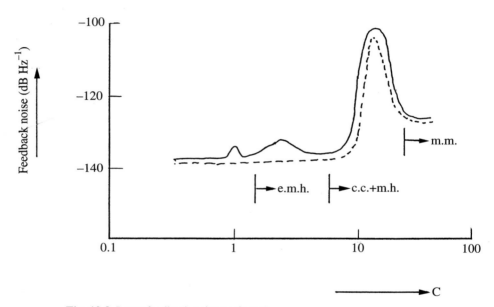

**Fig. 12.3** Laser feedback noise against *C*.

An alternative is to use optical methods to reduce the noise to a minimum. The half-aperture solution, is shown in Fig. 12.4. This divides the objective lens aperture into an incident-beam region and a reflected-beam region. This, however, reduces the numerical aperture. A more sophisticated solution is to impose an r.f. signal on the laser. The reflected light is then out of phase with this additional r.f. signal.

**Fig. 12.4** The half aperture solution to reducing laser feedback noise.

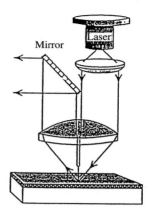

# 12.4 Media noise

## 12.4.1 Disk media noise

This noise source can be either intrinsic or extrinsic. The intrinsic noise can be due to the following:

- Polar Kerr angle variation due to composition changes in the M-O layer;
- reflectivity change of the M-O layer;
- refractive index changes and birefringence of the substrate;
- surface roughness in the band or groove;
- dust;
- poor thickness control greater than 5 % in the dielectric and M-O layers.

All of these intrinsic sources increase as the velocity of the disc is increased in order to produce drives with faster access times.

Extrinsic noise can be caused by disc motion noise, as the optical path changes owing to any erratic movement or wobble of the disc. This will obviously affect laser feedback noise also.

Light spot shape change when reading the data on the disc can also produce appreciable extrinsic disc media noise.

## 12.4.2. Writing noise.

When writing data on the disc there can be considerable noise generated by the variation in the shape of the written bits or domains in the media. Since the bits are vertically recorded through the film, then inter-symbol interference between bits on one track can be appreciable if the recorded regions are too close to each other. Track-to-track cross-talk is not generally a problem with track-to-track spacings in the range 1.3 to 1.6 $\mu$m.

Recorded data spot size can vary considerably if the laser power is not controlled sufficiently, since the temperature profile varies with the circumference in the direction of the rotation. An elongated domain can result, so that inter-symbol interference will be large.

Poor focussing will result in fuzzy edges to the domain and this can cause further noise.

## 12.5 Noise measurements

The noise and signal power in a 20 MHz frequency band is shown in Fig. 12.5. As the laser power at the disc required for reading is increased from 0.7 to 2.8 mW the output signal increases, as shown by curve G. The electronic noise excluding Shot noise, A, shows little variation with laser power. Shot noise, B, since it comes from light, depends on laser power. C = B+A. D is the total noise stopped; that is, when the disc is stationary and is mostly due to intrinsic media noise and laser feedback noise. E is the moving disc noise and now includes the extrinsic media noise in addition to the D noise sources. F is the total noise.

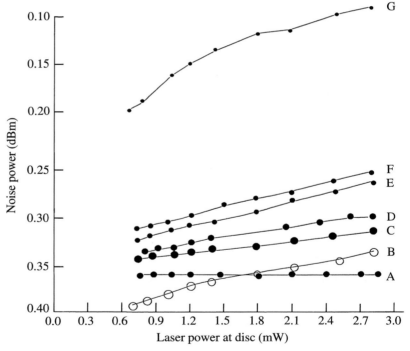

**Fig. 12.5** Variation of noise power with laser power at the disc. A, electronic noise; B, shot noise; C, shot + electronic noise; D, total noise when stopped; E, total noise when moving; F, total noise; G, signal level.

Moving on to Fig. 12.6, the variation of the noise sources with frequency is shown for a magneto-optic recording system. Only the Shot noise is a white noise source and shows no frequency dependence. The electronic noise, A, excluding Shot noise, increases with frequency. On the other hand the disc moving noise, D, the laser feedback noise, C, and the disc writing noise, E, are all higher at frequencies below 8 MHz and show little frequency dependence above this. For this particular system, the disc writing noise always dominates and the disc moving noise is always the smallest as the frequency of the signal increases.

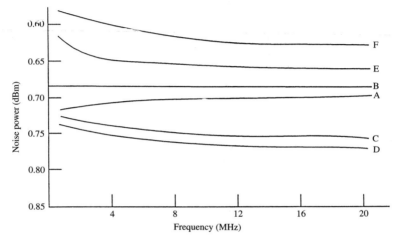

**Fig. 12.6** Variation of noise sources with frequency. A, electronic noise; B, shot noise; C, spurious noise (laser); D, disc noise when moving; E, disc writing noise; F, total noise.

Finally, the only noise source that has not been mentioned, which is very difficult to quantify, is owing to the residual signal after erasure. This is a serious problem, since with the first commercial system each track had to be erased three times to eliminate the remaining signal. Considerable effort has been devoted to developing solutions to overwrite problems for the second generation of optical recording systems.

**Question**

If disc writing noise is a major source of noise in magneto-optic drive systems, how can it be minimized?

**Answer**

Operate the laser at high frequencies of 8 MHz or above. Figure 12.6 shows the frequency variation of this noise source. The laser power must be carefully controlled, since this will affect the shape of written magneto-optic domain bit. The temperature profile during writing is obviously affected by the variation in laser power and by poor focusing of the light spot. The profile can be improved by using a sputtered layer with very carefully controlled thickness uniformity. Finally, amorphous materials like the rare-earth transition metal alloys have no noise caused by crystallite boundaries, as is found in cobalt-platinum multi-layers.

# Magneto-optical recording applications | 13

The greater part of this chapter will be concerned with computer applications. Four types of computer systems will be discussed:

- Portable
- PC-based
- Networks
- Mainframe

In every case there is a place for erasable magneto-optical recording. Some of the applications that are discussed are purely speculative, whereas others, like PC-based office document filing systems, are now well established.

Following this main section there will be a short one on sensors, since this is a new area where magneto-optical recording can be used to produce intelligent devices.

Finally, the future possibilities for hybrid systems are discussed, where multi-media combinations on one hard disc are used for educational and leisure purposes. Also to be considered will be the use of a hybrid silicon chip with a magneto-optic memory for a wide range of applications.

## 13.1 Portable systems

There are three types of portable systems which are listed below in order of decreasing size:

- Laptop
- Notebook
- Book

Currently the laptop market is dominated by the 3.5 inch (96 mm) fixed Winchester hard disc, but the removable 3.5 inch magneto-optic (or phase change) erasable disc will begin to replace the Winchester. Removable discs will be the key to this, since interchange of software and data with colleagues and clients is desirable.

Notebook-sized computers are also dominated by the fixed Winchester magnetic disc. To give one example of many, Prairie Tek have a 42.8 MB drive which weighs only 9.1 ounces (258 g) and it is 1.0 inch in height, 2.8 inches in width and 4.6 inches long ($25 \times 70 \times 115$ mm). Another key feature of the Prairie Tek drive is that the magnetic heads are parked at the side of the drive. This innovation means that the media life is extended to 250 000 start/stops.

In spite of these new developments the disc or discs are still fixed in the drive. The removable magneto-optic discs will enable a wide range of software to be used with this portable computer.

Finally, the book system is currently dominated by CD-ROM. To give one example, the portable Sony Data Discman, which began to be sold in Japan in 1990, can not only store about one hundred books on one 600 MB disc, but can also play CD audio discs. For this reason, the 'book' type of system, where erasability is not required, will continue to use CD-ROM. However, erasable magneto-optical systems could become important for the educational market where the student's individual course notes can be written on to blank discs.

## 13.2 PC-based systems

Document image processing (DIP) PC-based systems have been available in the UK since 1990. The Canon desktop electronic filing unit, Canon-file 250, which was probably the first system introduced in 1990, uses the Canon magneto-optic drive with an access time of 80 ms.

More recently, Sovereign has introduced a complete DIP system for just over £7000 (in 1992), which is well below the Canon price and offers a wider range of options. Four of the five basic components are shown in Fig. 13.1. The fifth is, of course, the software which is based on Windows 3 which supports a wide range of scanners and optical discs.

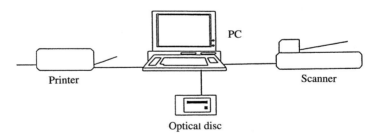

**Fig. 13.1** A document image processing (DIP) PC system.

It can operate with:

- 3.5 inch, 128 MB erasable disc;
- 5.25 inch, 600 MB 1 GB Maxoptics magneto-optical disc with an access time of 35 ms;
- 'Jukeboxes' with up to 100 GB of storage.

Various types of PC are offered: the cheapest is the 386SX, 20 MHz 2 MB memory, 40 MB hard disk, 1.4 MB floppy, 14-inch VGA mono display, with MS-DOS 5, Windows 3 and mouse; the most expensive is a 486DX, 33 MHz, 128 K cache, 8 MB memory, 80 MB hard disk, 1.4 MB floppy, 14-inch super VGA colour display, with MS-DOS 5, Windows 3 and mouse.

Business applications for these PC-based DIP systems include:

- Small manufacturing companies
- Banks
- Building societies
- Insurance companies
- Solicitors
- Garages
- Doctors' and dentists' surgeries

## 13.3 Networks

The first networked systems, which are still in use, were based on the Ethernet bus systems, as shown in Fig. 13.2. This uses either a WORM or a magneto-optical disc 'Jukebox' with a storage capacity of up to 530 GB. A large number of workstations can be accommodated, so this is particularly suited to large companies. Ethernet is limited to a data transmission rate of 10 MBs$^{-1}$, whereas the IBM Token Ring network system can go up to 16 MBs$^{-1}$.

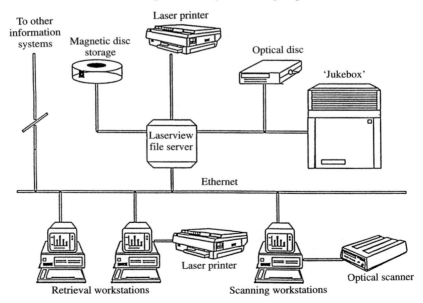

**Fig. 13.2** An Ethernet-based optical recording system for document management.

With the increasing introduction of fibre optics, network applications are expanding. FDDI (fibre distributed-data interface) networks have data rates of 200 MBs$^{-1}$ and as many as 500 computers can be connected up over a 100 km MAN (metropolitan area network). FDDI can also be used in small area LANs (local area networks) within a laboratory or office and is fully compatible with the 'hard-wire' Ethernet or IBM Token Ring systems.

## 13.4 Mainframe computer systems

Since 1998, mainframe data tape has been dominated by 3480 chromium dioxide tape with twice the track density of the open reel 3420 tape. It is an 18-track tape with a coercivity of 550 Oe and with a data density of 300 MB and 25,000 flux changes per inch (FCI). The data transfer rate is 3 MBs$^{-1}$.

RDAT is a much higher coercivity tape (1400 Oe with iron powder) and densities of 1.3 GB are possible with 61,000 to 76,250 FCI. Reed Solomon error correction is used, just as with all optical systems, and non-recoverable error rates of less than 1 in $10^{15}$ bits are possible. A MTBF (mean time before failure) of greater than 50,000 hours is claimed. The recording format is DDS as developed by Sony and Hewlett Packard. It has a standard SCSI interface, in the same way as optical drives. The average access time is 20 s and the synchronous burst transfer rate is 5.3 MBs$^{-1}$.

The bad news for DAT is that random file recovery can take up to 19 minutes, and on any tape that is used for long-term storage the data must be refreshed every two years. With the in-contact helican scan head technology, lubrication on the tape and head cleaning is also required.

In spite of these disadvantages, tape is accepted by mainframe managers, and as the Sony data storage pyramid shows in Fig. 13.3, it will continue to dominate mainframes even when the second generation of magneto-optic recording comes in.

**Fig. 13.3** The Sony data storage pyramid. HDD, hard disc data; FDD floppy disc data.

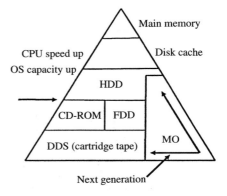

Within five years or so, however, erasable optical recording will begin to dominate as a back-up store because of its long-life advantage.

## 13.5 Major markets for computerized systems

The four major markets for magneto-optic disc recordings are in:

- CAD/CAM
- Mainframe back-up
- Military
- PC document file storage or DIP

CAD/CAM is particularly suited to erasable recording. Because the 5.25-inch drives are still being made in relatively low volume (250 000 installed by the end of 1991), the current high cost is not a barrier and with networking a central erasable store can be used. Moreover, the amount of data per drawing is increasing rapidly, so that a magneto-optic 'Jukebox' could be used for larger systems. Access times, which range from 35 to 100 ms, are all adequate for CAD/CAM.

The military market, as with CAD/CAM, sometimes needs very high security, so a fixed disc drive is preferred. In a fixed disc situation a number of discs can be stacked. This is referred to as the stacked optical drive, SOD. Hitachi have produce a prototype SOD with five rigid magneto-optical discs, and the outline specification is given in Table 1. The discs are 10 inches in diameter and double-sided, with a data capacity of 6GB per disc. The access time of 90 ms and total storage capacity of 30GB make the SOD very suitable for both military and CAD/CAM applications.

**Table 13.1**
**Outline specification of the stacked optical drive**

| Item | Specification |
|------|---------------|
| Capacity (per spindle) | 30 GB |
| Diameter (disc) | 254 mm |
| Transfer rate | 64 MBs$^{-1}$ |
| Channels | 10 |
| Disks (per spindle) | 5 |
| Rotation speed | 3600 rpm |
| Access time (average) | 90 ms |
| Track pitch | 1.4 $\mu$m |

Applications which will become increasingly important are the following:

- Medical
- Telecommunications
- Desktop publishing

Of these three, telecommunications will become by far the greatest. This will include banking, finance (stock market), teleconferencing, archival storage, accessing national and city directories and video security applications.

Medical and desktop publishing, although slow initially, will become increasingly important as the need for high density storage and resolution of pictures increases.

## 13.6  Magneto-optic sensors

Figure 13.4 shows that magnetic and magneto-optic sensors will have a place in the future of automation. In the case of the rotational magneto-optic sensor the device can be made intelligent. The scale can be changed as the tolerance increases or the number of degrees can be read optically to a very high accuracy.

**Fig. 13.4** Magnetic and magneto-optic sensors for automation.

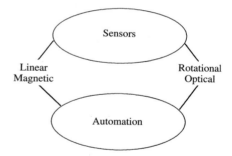

Linear magneto-optic sensors can also be made with built-in intelligence. They can be coded so that the direction and the positional accuracy can be read out.

Other new sensors are currently being patented, so these will be reviewed at a later date.

## 13.7  Further future possibilities

One development is the hybrid system where two or three types of media are combined on one optical hard disc. Sony refer to the partial ROM or the OD-

ROM, which has the outside of the disc as a CD-ROM and the centre portion has an erasable coating on it. This has obvious advantages, since commercial software can be 'printed' on the CD-ROM part of the disc cheaply and the magneto-optic part can be used for a wide range of memory applications. Fig. 13.5 shows the Sony concept of the OD-ROM. This same concept can also be used for multi-media systems in which sound, data and video are all combined.

**fig. 13.5** The rewritable partial ROM (CD-ROM disc size) including a rewritable portion and the full CD-ROM.

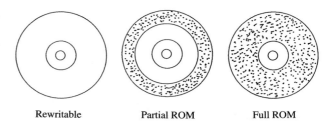

Rewritable          Partial ROM          Full ROM

One further possibility is to combine silicon chips with magneto-optic memories by depositing magneto-optic films on silicon slices. The silicon signal processing can be carried out on the front side while magneto-optic storage is achieved on the reverse side. Alternatively, by using photolithography, magneto-optic storage areas could be deposited on to silicon chips covering a wide range of sizes.

# Phase change media                                                14

Phase change media has two major advantages over magneto-optic media:

(1)  it can be read by reflectivity change alone;
(2)  direct overwrite of a single memory layer can be achieved by laser light modulation.

Consequently, considerable attention is being paid to them.

They do suffer from the disadvantage that the change in phase is only reversible for 100 000 times in commercial media, whereas 10 000 000 cycles has been achieved for magneto-optic media. However, in the research laboratories 2 000 000 write-read-erase cycles have now been achieved for phase change media, so further commercial development seems likely.

## 14.1 The phase change phenomena

Commercial phase change media was developed by Matsushita from the idea of Ovshinsky (1970) to use amorphous/cryatalline switching in optical recording. Out of a whole range of glassy materials the chalcogenides based on tellurium, selenium and sulphur were chosen. This was because an amorphous to crystalline phase transition temperature of between 100 and 200 degrees centrigrade was possible.

Marchant (1990) showed how the state of phase change materials depends on its thermal history. Figure 14.1 shows that if cooling occurs below $T_g$, the glass or phase transition temperature, no measurable change takes place. This means that reading with a low power laser, provided the temperature stays below $T_g$ will not alter the phase of the material, whether it is crystalline or amorphous.

Figure 14.1 also shows that quenching to the amorphous phase from the molten or liquid phase at temperatures above $T_m$, the melting point temperature, must be carried out rapidly. Annealing to the crystalline phase occurs above $T_g$. At high temperatures annealing occurs very rapidly. If the temperature is below $T_m$ and the heating is of short duration, then only partial recrystalization can occur in the heated zone.

Now, from generalities we move on to the specific material used most frequently by Matsushita: Germanium-antimony-tellurium. The composition diagram for this ternary system is shown in Fig. 14.2 according to Ohno, et al. (1991). Matsushita used compositions between point A and the point B. The composition at point A is: $Ge\,Sb_2\,Te_4$.

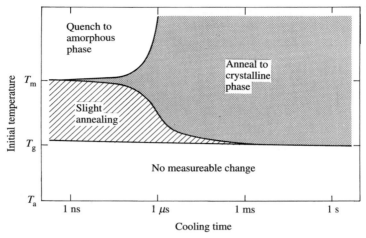

**Fig. 14.1** Temperature against cooling time showing how phase change material depends on thermal history. $T_a$, ambient temperature; $T_g$, glass transition temperature; $T_m$, melt point.

**Fig. 14.2** Composition diagram of the germanium, antimony, and tellurium ternary system.

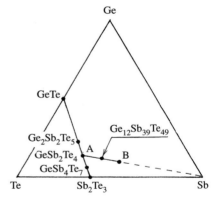

The composition sits exactly in the centre of the pseudobinary GeTe/$Sb_2Te_3$ phase diagram shown in Fig. 14.3 (Yamada, *et al.*, 1991). The vertical line in the centre at 50 % GeTe represents the stoichiometric compound $GeSb_2Te_4$. The properties of this compound and two others that are also found in this alloy system are shown in Table 14.1. This shows that, as one would expect, the phase transition temperature increases from 123 to 142 degrees Centigrade as the melting point increases. Referring back to Fig. 14.3, at the bottom of the diagram the phase transition temperature change is shown across most of the pseudobinary alloy system.

**Fig. 14.3** GeTe-Sb$_2$Te$_4$ pseudobinary phase diagram and phase transition temperatures.

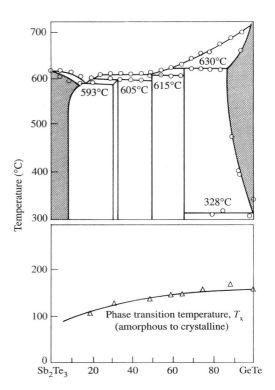

**Table 14.1**
Thermal analysis results

| Composition | Crystallization temperature(°C) | Activation energy (eV) | Melting point (°C) |
|---|---|---|---|
| Ge$_2$Sb$_2$Te$_5$ | 142 | 2.23 | 616 |
| GeSb$_2$Te$_4$ | 131 | 1.82 | 614 |
| GeSb$_4$Te$_7$ | 123 | 1.52 | 607 |

## 14.2 Disc construction for rapid cooling

The disc structure that has been developed for the Ge-Sb-Te films described above is shown in Fig. 14.4. This compares the conventional structure on the left with the rapid cooling structure on the right. The only difference is that the ZnS–SiO$_2$ dielectric top layer thickness has been reduced from 192 nm to 12 nm. The substrate is polycarbonate 1.2 mm thick and the overcoat is UV-resin, which is 50 $\mu$m thick.

The temperature profiles in the layers for the two types of structure are shown in Fig. 14.5. The profiles are at the instant when the peak laser power level is focused on to the Ge-Sb-Te layer. The laser spot is moving from the right to the left hand side at a speed of 8 ms$^{-1}$. For the conventional structure the typical overwrite power is: Pw/Pe = 16 mW/6 mW. The rapid cooling structure, on the other hand, has: Pw/Pe = 20 mW/10 mW.

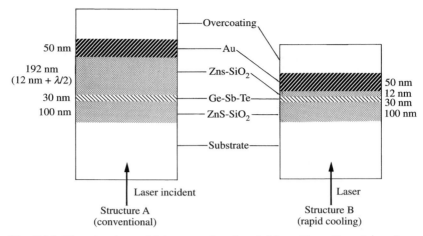

**Fig. 14.4** Disc structure of (a) conventional and (b) rapid cooling phase change media.

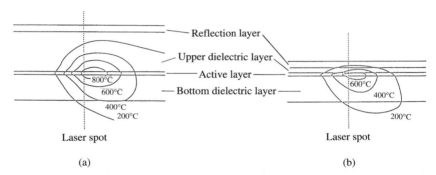

**Fig. 14.5** Temperature distribution in the layers shown in Fig. 14.4 by simulation calculations.

Fig. 14.5 suggests that:

(1)  the maximum temperature does not occur at the beam centre but to the right of it;
(2)  the conventional structure has to be heated to over 1000°C to obtain an optimum signal;
(3)  the improved disc structure shows a large temperature gradient in the vertical direction and this promotes rapid cooling.

(Note that the pulse width of the peak power is 60 ns for both structures.)

Finally, the carrier-to-noise (C/N) ratio and the erasability for these two structures are compared in Fig. 14.6; (a) is the conventional and (b) is the rapid cooling cooling disc structure. This clearly shows that a higher C/N ratio is achieved with the rapid cooling disc structure.

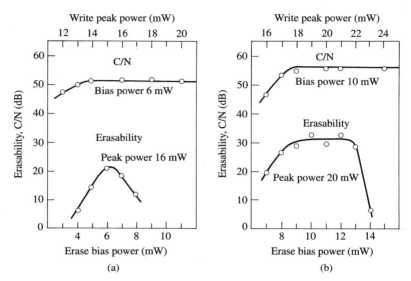

**Fig. 14.6** Laser power dependencies of signal-to-noise ratio and erasability for (a) thick dielectric and (b) rapid cooling disc structure. The linear velocity is 8 m/s and frequencies are $f_1 = 5.33$ MHz and $f_2 = 2.0$ MHz.

In both cases a 5.25-inch polycarbonate substrate disc was used. The disc speed was 2400 rpm and the linear velocity was 8 ms$^{-1}$ at the internal diameter tracks. Overwriting signal characteristics and erasability were measured using two different signal frequencies:

$F_1 = 5.33$ MHz equivalent to $1.5T$
$F_2 = 2.0$ MHz equivalent to $4T$ (where $T$ is the clock frequency).

The rapid cooling disc (b) shows an erasability of up to 30 dB and shows a constant erasability over the erase bias power range 8 mW to 13 mW.

## 14.3 Direct overwrite

Using the pulse shape or laser modulation, shown in Fig. 14.7, direct overwrite (OW) can be achieved. Figure 14.8 shows the reverse mode write technique in which crystalline material changes to amorphous when it is written (Ohara, *et al.*, 1990). Before overwrite, four amorphous or recorded spots are shown on the track below the pulse. When the erase laser power is applied this erases the amorphous spot by changing it to a crystalline state. This is because the erase bias power heats up the phase change media to a temperature between the crystallising and the melting temperature. The write power level raises the temperature to above the melting point and is quenched, so that an amorphous bit is written on the same track. Consequently, erasure occurs just before data is written with only one laser beam.

The shape of the recorded amorphous bit is normally pear-shaped. This can be improved by using the multi-pulse modulation method shown in Fig.

**Fig. 14.7** Laser power modulation scheme for single beam overwriting.

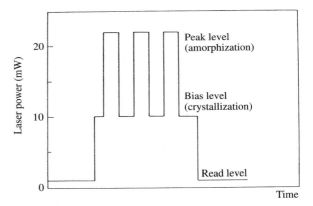

**Fig. 14.8** Direct overwriting showing a track before and after overwriting (OW). (Reverse mode write.)

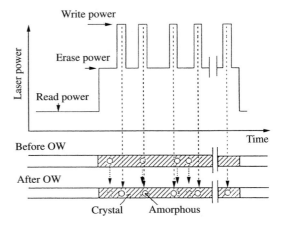

14.9 according to Nishiuchi *et al.* (1992). This shows the input digital signal, the laser output and the recorded mark, and compares (a), the multi-pulse modulation method with (b), the conventional method. Fig. 14.10 gives a more detailed picture of multi-pulse recording; (a) shows the input digital pulse of an 11T signal (where T is the clock period, equal to 23 ns); (b) is the 11T laser output showing a larger pulse of width W1 of 1.5T followed by eight smaller pulses of width W2 of 0.5T. (c) is the laser irradiation pattern on the disc and (d) the uniformly recorded mark.

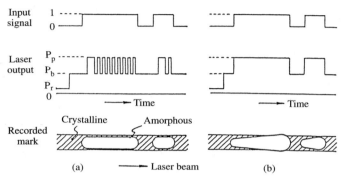

**Fig. 14.9** Schematic view of laser power modulation patterns and shapes of amorphous marks recorded at low linear velocity: (a) by a multipulse modulation method and (b) by a conventional modulation method.

**Fig. 14.10** Schematic view of multipulse recording: (a) original input waveform of 11T signal, (b) 11T multipulse signal, (c) laser irradiation pattern on the disc and (d) recorded mark.

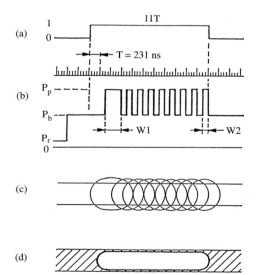

## 14.4 The phase change erasable drive system

Since the phase change is read optically by looking at the reflectivity change from the amorphous to the crystalline phase, it can be used in a WORM-type drive. A drive is shown in Fig. 14.11 (Ohara, *et al.*, 1990). The only difference with the WORM drive is that the laser drive circuit can control three power levels independently. As with the highest speed M-O drive a split optical head is used. The fine focus control actuator consists only of a mirror, and the two-dimensional actuator for focusing and tracking is moved by a common magnetic field (CMF) actuator.

The fixed part contains the laser (40 mW and 780 nm), polarising beamsplitter and a detector. 50 % of the laser power reaches the surface of the disc.

**Question**

What are the key requirements for phase change media?

**Answer**

(a)  A low glass or phase transition temperature in the range 150 to 200°C.
(b)  A large change in reflectivity (or refractive index) between the amorphous and crystalline phase. This enables 1's and 0's to be easily differentiated.
(c)  A low melting point which requires a smaller laser power.
(d)  Good alloy homogeneity so no phase separation occurs after melting.
(e)  Expansion coefficients of the crystalline and amorphous phase must be similar, or cracking will occur.
(f)  Rapid annealing below the melting point results in single-pass erasure and no overwrite problems.

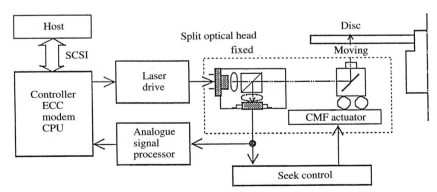

**Fig. 14.11** Schematic configuration of the phase-change optical disc memory showing laser drive control.

Table 14.2 gives the specification of the Panasonic multifunction phase change rewritable 130 mm WORM disc drive LF7010E. The average seek or access time is still much slower than the best M-O drive, but the capacity of 1 GB is the same. The modulation coding is RLL(2–7).

Experimental drives have been reported by Ohara, *et al.* (1990) with an access time of 42 ms for an 86 mm disc drive with a disc speed of 2400 rpm and a disc capacity of 280 MB equal to that of the M-O.

**Table 14.2**
**Panasonic LF-7010E multifunction drive**

| Type | Phase change | WORM |
|------|--------------|------|
| Average Seek | Less than 90 ms | |
| Data transfer rate (to/from disc) | Average 7.92 MBs$^{-1}$ | |
| Motor start time | 5 s | |
| Disc speed | 1800 rpm | |
| Corrected bit error rate | Less than $10^{-12}$ | |
| Interface | SCS12 | |
| MTBF | 20 000 hours | |
| Disc diameter | 130 mm | |
| Disc capacity (2 sides) | 1 GB | 940 MB |
| Sectors per track | 17–32 | 18–32 |
| Sector size | 1024 bytes | 1024 bytes |
| Tracks per side | 19 968 | 18 360 |
| Track pitch | 1.5 μm | 1.5 μm |

# 14.5  Life measurements

Two types of measurements are reported here: write, read and erase cycles; and accelerated life tests.

First, the write, read and erase cycle test will be looked at. Figure 14.12 shows the reflectivity change for recorded amorphous regions and crystallised erased regions of $GeSb_2Te_4$, with a record laser power of 25 mW and erase laser power of 12 mW for a 40 ns pulse. This was a static test, and it shows that some changes in reflectivity do occur. However, after $10^6$ cycles the reflectivity difference between the two phases was still easily distinguishable.

A carrier-to-noise ratio of 50 dB has been achieved for $2 \times 10^6$ overwrite cycles with a moving disc (Ohta, *et al.*, 1989). (The alloy came from the GeSbTe system.)

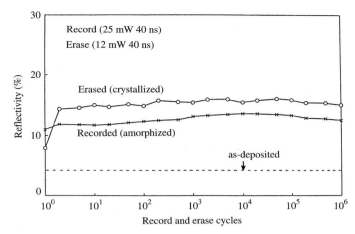

**Fig. 14.12** Static record/erase cycle test results for $GeSb_2Te_4$ thin film showing good stability in reflectivity properties.

## 14.6 Conclusion

Phase change media, although commercially available, is still not fully accepted. This is because the life problems that it initially suffered from have still not quite been fully overcome. This has allowed MO media to achieve acceptance and this will be hard to change.

Multifunction drives are possible since both WORM and phase change erasable media use reflectivity change. However, since WORM has a limited market, this is not going to help phase change to become a major competitor to the hard magnetic Winchester disc.

### References

Akahira, N., Yamada, N., Kimura, K., and Takao, M. (1988) *Society of Photo-optical Instrumentation Engineers*, **899**, pp. 188–195.

Marchant, A. B. (1990) *Optical Recording*, Addison-Wesley, p. 86.

Nishiuchi, K., Akahara, N., Ohno, E., and Yamada, N. (1992) *Japanese Journal of Applied Physics*, **31**, pp. 653–8.

Ohara, S., Fukushima, Y., Miyazaki, B., Moriya, M., Satoh, I., and Yoshida, T. (1990) *Society of Photo-optical Instrumentation Engineers*, **1248**, pp. 74–85.

Ohata, T., Furukawa, S., Yoshioka, K., Uchida, M., Inoue, K., Akiyama, T., Nagata, K., and Nakamura, S. (1990) *Society of Photo-optical Instrumentation Engineers*, **1316**, pp. 367–373.

Ohno, E., Nishiuchi, K., Yamada, N., and Akahira, N. (1991) Optical data recording, *Society of Photo-optical Instrumentation Engineers*, **1499**, pp. 171–9.

Ohta, T., Uchida, M., Yoshioka, K., Inoue, K., Akiyama, T., Furukawa, S., Kotera, K., and Nakamura, S. (1989) *Society of Photo-optical Instrumentation Engineers*, **1078**, pp. 27–34.

Ovshinsky, S. (1970) US patent no. 3530441.

The success of optical data recording is assured because:

(1)  the optical reading head is 1 mm away from the disc and head crashes do not occur;
(2)  the media is removable and can be used to interchange a wide range of information;
(3)  the media price per MB is now below the cost of a floppy disc;
(4)  hybrid media such as the erasable CD-ROM, in which a read only memory is combined with an erasable section, is suited to multimedia applications in which video, data and audio can all be combined;
(5)  further technical improvements are currently underway, so that 10 GB of storage can be achieved on a 130 mm disc.

Technical improvements that are currently under development include the following:

- Laser wavelength – move to 530 mm
- Numerical sperture of the objective lens
- Pit or domain-edge recording and reading
- MCAV format

A new development by AT&T Bell Laboratories uses a tapered optical fibre with a flat tip 20 nm in diameter. Bits 60 nm in diameter and track-to-track spacing of 120 nm have been achieved on conventional magneto-optic media. Densities of close to 7 GB cm$^{-2}$ have been reported. One main problem with this breakthrough in technology is that of constructing a cheap drive which can read such small domains. The other is that of the fibre tip which will have to fly close to the surface – just like a magnetic head.

Finally, the challenge that was mounted by the magnetic recording industry in the late 1980s led to some impressive gains in media recording densities and access times in the Winchester drive. Faster access times are now being achieved by optical drives, but the acceptance of the CD-ROM in the early 1990s has accelerated the introduction of optical recording. Multimedia, when the standards settle down, will ensure that optical recording will gradually replace magnetic tape and floppy discs.

# Glossary

**aberration**: phase distortions which degrade the performance of an optical system. Aberrations can be caused by defective optical elements, defocus, misalignment, etc.

**access time**: the time to get to or from a specified location in or on a memory device. For disc drives, quoted access times may refer only to the positioning time for the radial actuator, neglecting servo settling and rotational latency. Usually quoted in ms.

**actuator**: a device for moving or actuating a payload in response to a control signal; can work open loop or as part of a servo loop.

**ADPCM**: Adaptive Pulse Code Modulation. A process for storing audio information.

**address**: the physical or logical location in a computer's memory of a stored quantity of data.

**air-sandwich disc**: an optical disc composed of a substrate bonded at its inner and outer edges to another rigid part (usually an identical substrate), with a thin-air space covering the recording zone. Air-sandwich discs usually operate in a substrate-incident mode and are usually designed to be double-sided.

**amorphous**: having no form; more specifically, having atoms arranged at random. An amorphous substance softens gradually and, as a result, has no single melting point. Some erasable optical media use materials which can be flipped back and forth between amorphous and crystalline states.

**angstrom**: a unit of length used in measuring light waves, etc.; one ten billionth ($10^{-10}$) of a meter, abbreviated A or Å.

**anode**: a terminal, realized as a structure such as a plate, wire, or grid, into which electrons flow; a positive electrode.

**antireflection bilayer, trilayer, and quadrilayer**: media designs that include reflective layers and/or transparent layers that enhance the absorption of light in the recording layer. Antireflection designs increase media sensitivity and signal contrast.

**archival**: media which is readable (and sometimes writeable) for a long time; 'long' means anywhere from five years to more than 100 years.

**ASCII**: American Standard Code for Information Interchange; a standard table of seven bit designations for digital representation of upper and lower case Roman letters, numbers, and special control characters in teletype, computer, and word processor systems. ASCII is used for alphanumeric communication by everyone except IBM, whose own similar code is called EBCDIC. Since most computer systems use a full byte to send an ASCII character, many hardware and software companies have made their own non-standard and mutually incompatible extensions of the official ASCII 128 character set to a 256 character set.

**astigmatism**: an aberration in which a beam of light converges not into a round spot, but into two perpendicular lines spaced some distance apart along the length of the optical path. The distance between the two lines is a measure of the astigmatism.

**ATL**: abbreviation for automated tape library.

**axial**: in a direction parallel to the axis of a rotating part; for a disc perpendicular to the plane of the disc.

**axial acceleration**: for a disc, the acceleration of the disc's surface in the axial direction.

**axial runout**: the peak-to-peak motion of the recording layer of a spinning optical disc in the direction parallel to the rotation axis.

**azimuth**: an angle on the horizontal plane, as distinct from an angle of elevation. Used to describe location along a circular track.

**B**: abbreviation for byte.

**back-up**: a copy of stored data; e.g., data stored on a fixed magnetic disc, made to prevent catastrophic loss of data in case of drive failure.

**band-rate**: used in data communications to define the rate of data bits per second.

**baseline reflectivity**: the reflectivity of an unwritten part of a writeable medium.

**BER**: bit error rate. BER is expressed as the number of data bits encountered before one erroneous bit is found (e.g. one erroneous bit in $10^{12}$ bits).

**bias field**: an external magnetic field applied normal to the surface of an M-O medium during recording and erasure.

**bilayer**: regarding discs, one whose recording structure contains two layers (e.g., a reflector layer and a dye layer).

**birefringence**: an effect in which the refractive index of a material is different for different polarizations of light, leading to aberration of a beam traversing that material.

**bit**: abbreviation for binary digit; the minimum unit of data information expressed in binary form; that is a 1 or a 0. See also 'byte'.

**bit error rate**: see **BER**.

**block**: an amount of data moved or addressed as a single unit; the least amount of data to be read or written at a time. Typical block sizes in optical disc drives are 0.5 kB to 2 kB.

**block code**: an error-correcting code in which a string of additional bits is appended to a block of incoming data; see **convolutional code**.

**bpi**: abbreviation for bits per inch.

**broadband**: a general term referring to a wide range of frequencies, a device that can accept a wide range of frequencies, etc.

**broadband absorption**: absorption of a wide spectrum of (laser light) wavelengths. A medium with broadband absorption would be usable with a greater number of different lasers than a medium capable of absorbing only a limited range of wavelengths.

**bubble**: in optical memory, those bubbles which are formed by a laser in an optical recording medium. Where bubbles exist, light is reflected away from the data reading optics, thereby creating a contrast to the bubble free areas of the medium surface. CF. 'hole', 'pit', 'mark', 'crater'.

**buffer**: 1 (noun) a relatively small portion of memory in which data is kept briefly between or during steps of processing of that data. 2. (verb) the use of a buffer to store data. Examples of applications of buffering include matching sending and receiving rates of different pieces of equipment and holding some data which is likely to be used in the very near future.

**burst noise**: in error correction, the loss of many consecutive bits of information, usually because of some flaw in the medium such as a scratch or dirt.

**byte**: a unit of digital data consisting of eight bits. Frequently called a 'character' by computer manufacturers describing capacities of memory devices, on the grounds that most computer systems will devote a full byte to storing an ASCII character. Abbreviated B. See also **bit**.

**cache memory**: generally temporary storage for data to which access must be very quick for the purpose of increasing computerized system operating speed.

**CAR**: computer-assisted retrieval. Broadly, using the search capability of the computer to retrieve the desired information, which may be stored on micrographics or some other mass media.

**carrier**: an electromagnetic wave; e.g., light wave, radio wave, microwave, which is modulated and therefore carries information.

**carrier-to-noise ratio**: abbreviated 'CNR'; the ratio of the carrier level to the level of noise also found in the channel of interest. CNR results for optical storage media re usually stated with respect to a 30 kHz bandwidth, which is much narrower than the bands usually used to carry video or digital information; as a result, CNR has value mostly as an arbitrary figure of merit with which everyone is familiar.

**case**: a protective container in which the optical disc is permanently resident.

**cathode**: 1. a terminal, realized as a structure such as a plate, wire, or grid, out of which electrons flow; a negative electrode. 2. a disc replication, the role of the metallized master disc in the galvanic copying process, in which nickel is plated onto the master in order to make a 'father' or master negative.

**CAV**: constant angular velocity. Describes a disc which always spins at the same rotational rate, so that the time taken to scan a track is the same at all radii, see 'CLV'.

**CCD**: charge coupled device. A semiconductor device capable of both photodetection and memory, which converts light to electronic impulses. One and two dimensional CCD arrays are used in scanners to perform the first stage in converting an image into digital data.

**CD**: Compact Disc. The trademarked name for the laser-read digital audio disk, 12 cm in diameter, developed jointly by Philips and Sony; see 'CD-ROM'.

**CD-I**: Compact Disc-Interactive. A multimedia standard developed by Sony and Philips for storing and retrieving text, audio, graphics and video information on Compact Disc.

**CD-ROM**: Compact Disc Read Only Memory; a version of the standard Compact Disc intended to store general purpose digital data; provides 556 MB user capacity at $10^{-13}$ corrected BER compared to 635 MB at $10^{-9}$ for the standard CD.

**CD-ROM cartridge**: a lockable plastic case to hold and protect a CD-ROM which may be inserted into the CD-ROM player. Also known as caddy.

**CD-ROM drive**: hardware for retrieving (reading) data on Compact Disc for use with PCs.

**channel code**: a code which represents information as a signal which can pass through a given channel; examples include the FM code used to put video on videodisks, the EFM code used to put digital data on Compact Discs, and the (2,7) run length limited code used to put digital data on some writeable optical discs.

**CIRC**: abbreviation for Cross-Interleaved Reed-Solomon Code, the error correction code used in the Compact Disc.

**circumferential**: running along a circle, for instance, moving over a track in a disc.

**clamping zone**: the part of a disc which is in contact with the spindle and spindle clamp.

**clip-art**: an archive of images stored in CD-ROM or floppy disc form for use in desktop publishing.

**CLV**: constant linear velocity. Describes a disc which turns more slowly when outer radii are being scanned so that the relative velocity between the light spot and the track is maintained at a constant value. This keeps the linear density of data constant over the whole disc, but creates practical problems due to the non-stant time taken to scan one track and the need to speed up and slow down the disc as various radii are scanned; see 'CAV'.

**CNR**: abbreviation of carrier-to-noise ratio.

**coating**: material applied in one or more layers to the surface of an optical element to change the way it reflects or transmits light.

**code**: a method or formula for representing information; see 'channel code' and 'error correcting code'.

**coercivity**: the strength of a magnetic field required to change or switch the existing magnetic state of a material. Magnetic materials lose coercivity as the temperature rises; see 'Curie temperature', 'compensation temperature'.

**coherence**: the degree to which the waves in a beam of light are in a stable phase relationship. The notable characteristic of laser light is its high coherence, which allows it to be focused to a small, intense spot.

**collimate**: the art of making light rays parallel.

**collimator**: a device, such as a lens, which turns an available light beam into a parallel (collimated) beam.

**Compact Disc^tm**: trade name for the 12 cm (4.72 inch) consumer optical read-only digital audio disc; also known as CD; see 'CD', 'CD-ROM'.

**compensation temperature**: the temperature at which the magnetization of a ferrimagnetic compound temporarily drops to zero; also known as 'compensation point'. Magneto-optic materials which are ferrimagnetic are often written by bringing them locally up to the compensation point in the presence of a localized magnetic field.

**compression**: in the specific context of digital image representation, refers to the process of compacting the data based on the presence of large white or black areas in common business documents, printed pages, and engineering drawings. The CCITT digital facsimile standards contain standard one and two dimensional compression /decompression algorithms.

**concentricity**: the radial distance between the best-fit center of a track and the rotation axis or the center of the disc's inner diameter.

**controller**: a specialized computer or device used to control the flow of data between a computer and one or more memory devices, usually a tape or disc drives. Controllers sometimes also perform channel and error correction coding and decoding.

**convolutional code**: an error correction code which continuously transforms the incoming data sequence into a longer encoded sequence; see 'block code'.

**crater**: a mark made by transverse displacement of the material of an optical recording medium. A crater is distinct from a hole in that a crater has a raised 'lip' around its perimeter; see 'hole', 'mark', 'pit', 'bubble'.

**cross-talk**: interference in the readout waveform due to neighboring tracks, which may be partially visible to the optical stylus. Erasure cross-talk results when a previously written track is not completely erased. Servo cross-talk is the tendency for tracking errors to produce spurious focus-error signals and vice versa.

**Curie point**: same as 'Curie temperature'.

**Curie point recording**: writing, for example on a magneto-optic disc, by bringing the temperature locally up to the Curie point, so that a weak magnetic field can magnetize the heated spot.

**Curie temperature**: the temperature above which a material permanently loses any magnetization it had. The temperature at which coercivity drops to zero.

**DAD**: abbreviation for digital audio disc; usually refers to Compact Disc

**D/A**: digital-to-analogue. Refers to the conversion of digital data to analogue signals.

**Data compression**: a method of reducing the space data uses whilst stored in memory devices which can subsequently be decompressed as required for reading.

**Depletion layer**: an electron barrier at the junction of a metal and a semiconductor due to depletion of charge carriers.

**diffraction**: the scattering or spreading of light waves.

**diffraction limit**: the smallest optical spot that can be formed by an optical system in the absence of aberration.

**divergence**: the spreading of a light beam with distance.

**DOS**: disc operating system. A major function of the central processing of a computer system which provides access to customized programmes and data held on disc or other forms of media. See 'operating system'.

**DRAW**: **direct-read-after-write**: optical media in which the marks are visible immediately after the recording exposure (say, within 1 $\mu$s.) DRAW systems exploit this characteristic to identify any hard errors as soon as they are written.

**DVI**: Digital Video Interactive: a form of data compression used mainly for digital audio and video information.

**E-beam**: short for electron beam, a popular tool for evaporating materials for high volume vacuum deposition of thin films.

**ECC, error-correction code**: an encoding scheme that appends generalized bits to a block of data before recording, and then uses the redundant bits to locate and correct errors after the data are retrieved.

**EDAC**: abbreviation for error detection and correction. This includes all phases of identifying and dealing with data errors, including direct-read-after-write and error correction codes.

**EFM**: The channel code used in Compact Disc, in which each 'word' of eight data bits is turned into fourteen channel bits; see 'channel code'.

**electromagnetic radiation**: waves made up of oscillating electrical and magnetic field perpendicular to one another, travelling at the speed of light. Waves can also be viewed as photons or quanta of energy. Electromagnetic radiation includes radio waves, microwaves, infrared waves, visible light, ultraviolet radiation, X-rays and gamma rays. (J. Hecht).

**erasable**: media which is capable of being rewritten, either after bulk erasure or spot erasure; see 'magneto'optic'.

**error burst**: a string of bit errors.

**error correction code**: a code which turns a bit stream into a longer bit stream whose length comes from carefully designed redundancy. This is intended to enable the encoded bit stream to survive corruption by random noise and burst noise and still be decodable to the original bit stream without missing or wrong data; one example is CIRC, the cross interleaved Reed-Solomon code used in the Compact Disc system.

**error detection and correction**: see 'EDAC'.

**Faraday effect**: certain substances, when exposed to a magnetic field, will change the polarity of light passing through them. This polarity change is called the Faraday effect. Optical discs using the Faraday effect must be transmittive, rather than reflective.

**feedback loop**: a circuit in which part of the output signal is used to modify the input signal for control purposes. For example, a simple feedback loop consists of your hand, brain, and eye whenever you pick up a glass of water. Your eye sends a signal to the brain. This input signal tells the distance and direction between glass and hand. That information controls an output signal that goes from brain to hand, causing the hand to move closer to the glass, which, in turn, causes the eye to feed back a new control input signal to the brain. Feedback loops also control focus and tracking of optical discs by laser beams; see 'focus servo', 'tracking servo', 'servomechanism'.

**ferrimagnetic**: weakly attracted by a magnetic field.

**ferromagnetic**: strongly attracted by a magnetic field.

**fill factor**: the degree to which fluid plastic fills in all the detail in a mould; affects signal quality in replicated discs.

**form surface**: the outer dimensional constraints for a standard-sized drive. The description of a particular form factor (e.g., 5 1/4″) usually bears a loose, historical connection to the actual dimensions.

**front surface**: the first surface on which light impinges. In front-surface readout, the illumination is air-incident, not focused through a transparent substrate.

**GaAs**: gallium arsenide, used in LED's, laser diodes, and other semiconductor compounds.

**Green Book**: standards which define the characteristics of CD-I discs and drives.

**groove**: a continuous channel cut or moulded into a surface to provide guidance to an optical head. Optical media made with patterns of moulded or photo-etched grooves are called 'pregrooved'.

**GB**: gigabyte, or 1000 MB.

**Guard space**: see 'error burst'.

**handling zone**: the part of a disc that may be touched by a handling mechanisms; e.g., in a jukebox.

**-H loop**: the plot of applied versus residual (remanent) magnetic field which characterises a magnetic material; the characteristic loop form of such a plot shows the hysteresis, or memory, characteristic of such materials and is basic to magnetic and magneto-optic storage.

**heterodyne**: generally, to change the frequency of a signal. Specifically, a method of sending video colour information on a lower subcarrier frequency than in broadcast television.

**hole**: a mark burned through the information bearing layer of an optical medium; distinct from a crater in that a hole has no raised 'lip' around its perimeter; see 'bubble', 'pit', 'mark', 'crater'.

**holographic lens**: a hologram of an object is a device which when illuminated by a laser, recreates the light which had bounced off of or passed through the object, which is what a person sees when observing a real object. Therefore a hologram appears to reconstruct the original object, complete and in three dimensions. If the object recorded on a hologram is a focused spot of light, subsequent illumination of the hologram recreates the spot of light. This sort of hologram thus functions as a lens.

**Imbalance**: a measure of the assymetry of the distribution of material in an object; rotational imbalance in a disc correlates directly to the disc's tendency to wobble when spun.

**index of refraction**: the amount that light bends when it enters a material. Properly, the ratio of the speed of light in a vacuum to the speed of light in a material. For example, light travels about 2/3 as fast through glass as through a vacuum, so the index of refraction of glass is 1.5.

**interactive video**: a combination of hardware, software and data in which the user determines the process or route through the available information. Published IV titles are available in CD-ROM and 12-inch laserdisc form.

**interface card**: computer hardware that plugs into a PC to allow the disc operating system to communicate with CD-ROM drives or other peripherals.

**interference**: the variation caused by the superposition of two or more light or electromagnetic waves. In optics, really in a series of dark and light bands or friezes which may be used for accurate measurements. See 'interferometry'.

**IS08855/1**: an International Standards Organisation specification for CD-I formats.

**IS09660**: an International Standards Organisation specification for CD-ROM formats.

**juke box**: a CD-ROM player housing 50 to 200 or more 12-cm CDs.

**Kerr rotation**: a magneto-optic effect in which the polarization of a light beam is rotated slightly upon reflection from a magnetized layer. The direction of polarization rotation depends on whether the magnetization is predominantly parallel or antiparallel to the beam incidence direction. Typical M-O films exhibit Kerr rotation angles less than 1°.

**land area**: on pregrooved substrates, any area on the surface which corresponds to unexposed portions of the photoresist master. For write-once and erasable media, the land area further excludes written marks or pits.

**LASER**: acronym for Light Amplification by Stimulated Emission of Radiation used as the coherent invisible light source in all CD and CD-ROM drives.

**laser diode**: variously called a diode laser, semiconductor laser, or injection laser. Laser diodes used for optical recording are driven by direct current injection and emit light at near-infrared wave-lengths.

**laser feedback**: a condition in which light is reflected back into the cavity of a laser. Feedback can alter a laser's efficiency, spectrum, power output, and noise.

**laser isolation**: any optical technique that prevents laser feedback.

**laser yield**: in manufacturing lasers, the percentage of finished product that meets specified requirements.

**latency**: that component of the delay in access to data which comes from waiting for a disc to rotate to the desired azimuth. Average latency for a disc drive is usually one-half the rotational period.

**light pipe**: concentration of light in a laser junction at a region of low electron density.

**loop**: a circuit of which some of the output is put into the input; see 'closed loop', 'feedback loop', 'open loop', 'servomechanism'.

**loop gain**: in a servomechanism, the ratio by which offsets presented to the input are reduced by the active loop. For example, the focus servo of a typical videodisc player has a low frequency loop gain of more than 60 dB, meaning that a one millimeter deflection of the disc will result in only a one micrometer focus error.

**MB** megabyte–one million bytes of digital information or storage space. Maximum capacity of a CD-ROM is approximately 680 MB.

**magneto-optic**: information stored by local magnetization of a magnetic medium, using a focused light beam to produce local heating and consequent reduction of coercivity so that a moderately strong, poorly localized magnetic field can flip the magnetic state of a small region of high coercivity material. Reading is done either magnetically, with inductive heads in close proximity to the medium, or optically, through sensing the rotation of the plane of polarization of probing light via the Faraday effect or Kerr effect.

**mainframe**: a large, expensive, powerful computer intended for centralized application.

**mark or pit**: a region with altered optical characteristics created by a high-power exposure from the optical stylus. The visible mark is called a pit when it corresponds to ablation or flow of material in the recording layer.

**mark geometry**: the size and shape of the mark made by a laser on an optical medium.

**master**: an original recording, in its final form, intended for mass replication, often made from glass from which a stamper is produced for creating working copies.

**MCLV**: modified constant linear velocity. In MCLV, tracks are divided into bands. Within a band, the disc spins at a constant angular velocity, but that velocity is different for each band. The relation between velocity and band location is similar to the velocity versus radius curve for CLV operation.

**mechanically accessible zone**: the ring shaped zone on and around a disc to which the optical head has unobstructed access during operation.

**media**: properly, plural of 'medium', but widely used as both singular and plural.

**medium**: a substance or object on which information is recorded and stored; refers either to the sensitive coating on a writeable device or to the device itself; e.g., magnetic disc, tape, card etc. Usually refers to removable media, not fixed as in hard discs.

**micrometer**: one millionth of a meter, or about .00004 inch.

**micron**: old name for micrometer.

**mirror area**: a land area that is not adjacent to grooves and contains no data marks or pits.

**M-O, magneto-optics**: refers to methods or materials in which magnetic fields are made 'visible'. In M-O recording, data are stored as magnetic domains and recovered optically, usually using the Kerr rotation effect.

**multi-media**: CD-ROM applications embracing text, audio, video and graphics.

**numerical aperture**: the sine of the half-angle of the convergent code of light entering or emerging from a lens; abbreviated NA. Practically, a measure of optical performance: the higher the NA, the smaller the light spot. However, high NAs lead to tight tolerances on defocus, window thickness, and other important parameters.

**OCR**: abbreviation for optical character recognition. The ability of a computer to recognize written characters through some optical sensing device and pattern recognition software.

**ODD**: optical data disc.

**Oe**: oersted, the unit of magnetic field strength, coercivity.

**OEM**: original equipment manufacturer. Usually refers not to the manufacturer of a device, but to the system integrator who assembles and resells the device as part of a system. Sometimes used as a verb, as in 'Company B is going to OEM company A's drive'; this means that company A will manufacture the drive and company B will integrate it into a system to sell on.

**oersted**: the unit of measurement of coercivity, magnetic field strength.

**OMDR**: optical memory disc recorder.

**open loop**: the condition in which a circuit intended to act as a feedback loop has the line bringing the feedback disconnected from the input; usually done for test or measurement. Also used to describe the circumstances obtaining before the feedback is applied.

**operating system**: a specialized program which provides a computer with its basic 'personality', including its interfaces with human users and peripheral devices such as printers, displays, discs and memories (see 'access method'). Some microcomputer operating systems are CP/M, MS-DOS and Windows. One for larger microcomputers and minicomputers is UNIX.

**optical beam accessible zone**: the ring shaped zone on an optical disc to which the read or write beam has unobstructed access during operation. It is included with the 'mechanically accessible zone'.

**optical card**: a rectangular optical recording medium on which the data are not arranged in circular tracks. The credit-card-sized Drexon® Laser Card is the protypical optical card.

**optical disc**: a disc read and/or written by light, generally laser light; such a disc may store video, audio, or digital data.

**optical interferometry**: refers to the use of wave interference in light (see 'interference') to measure very small distances, height variations, variations in refractive index of a material, etc. Expertise in interferometry is important in designing certain kinds of media and in practising quality control or other media diagnostics.

**optical path length**: a useful measurement of the distance light travels through a medium. Technically, it is derived by multiplying the thickness of the medium by the medium's index of refraction. With an optical disc, the optical path length is usually measured between the point at which the laser beam first enters the disc and the sensitive layer, rather than the thickness of the whole disc. If the disc has more than one layer, the optical path is derived for each layer, and these are added together to derive the disc's optical path.

**optical stylus**: the spot formed by focusing a laser beam through a diffraction-limited objective lens which is used to record and read marks on optical recording media.

**OSI**: open systems interconnect. A mass storage interface standard promulgated by the ISO.

**outrigger tracking**: a technique for generating a tracking-error signal by comparing the signal levels from two low-power focused spots on either side of the optical stylus.

**overcoat**: a transparent protective layer for optical recording media, intended to keep dust and scratches out of contact with, and out of focus relative to, the actual information bearing layer. Overcoats are typically cast or otherwise coated directly on to the information surface; these are 'in contact' with that surface.

**overhead**: in computer jargon, the amount of storage or other resources used to accomplish tasks other than storage and transmission of data. For example, error correction code added to data might increase total storage requirements by 20%, which would be referred to as the overhead of that error correction code. Also applies to addressing and formatting.

**PCM**: see pulse code modulation.

**phase-change**: an optical recording process in which the recording layer is changed from crystalline to amorphous, or vice versa. In some materials, a large change in reflectance accompanies the chance of phase.

**photodetector**: a solid-state electronic device that converts light intensity into a corresponding electrical current.

**pit**: broadly used to refer to data-carrying marks in optical media, but originally coinced to describe the rimless troughs written in photoresist on videodisc masters and transferred by moulding to videodiscs, see 'hole', 'crater', 'bubble', 'mark'.

**pitch**: the distance between tracks (track pitch) or marks (bit pitch).

**pixel**: abbreviation for 'picture element', the smallest resolvable dot in an image display, especially TV or monitor displays.

**PMMA**: polymethylmethacrylate, a rigid, transparent acrylic plastic sold under the trade names of Perspex, Lucite, and Plexiglas. Many LaserVision videodiscs are injection moulded from PMMA. It tends to absorb moisture but otherwise has excellent media characteristics.

**polarity**: the sense of a charge, terminal, pole orientation, or region e.g., positive or negative, north or south.

**polarization**: a condition of the displacement between the electric and magnetic field vectors during propagation of light or electromagnetic waves. Normally, if a beam of light is viewed head on, so that its cross section were like the face of a clock, light waves would be oriented at every second around the dial: 12 o'clock to 6 o'clock, 1 to 7, 2 to 8, etc. This is unpolarized light. If light is plane or linearly polarized, only one orientation of waves is found; e.g., only 3 o'clock to 9 o'clock is present. The light from a laser is usually linearly polarized. The quarter wave plate found in many optical heads turns this into circular polarization, in which two perpendicular plane-polarized components are created with a 90 degree relative phase. Other combinations of phases and amplitudes of two perpendicular components are described as elliptical polarization.

**polarizer**: a device that transmits light of only one polarization.

**polycarbonate**: a rigid, transparent plastic, strongly resistant to environmental and handling effects. Most Compact Discs are compression moulded from polycarbonate.

**polyimide**: (pa-lee-im-ide) a kind of plastic composed of long chains of imide molecules.

**polymer**: a material composed of long chains of molecules. Most plastics are polymers. The names of specific polymers usually consist of the prefix poly- and the name of the base molecule; e.g., polyimide, polycarbonate, polymethylmethacrylate.

**polymethylmethacrylate**: 'PMMA' used in manufacture of optical discs.

**polystyrene**: a sturdy plastic considered for use as a substrate material for optical discs.

**pre-grooving**: the practice of moulding, casting, or otherwise producing on the substrate of an optical storage medium a physical guidance pattern which can be found by the tracking servo of the drive for which the medium is intended.

**pre-mastering**: the data conversion stage prior to mastering which combines error-detection codes with raw data.

**protective layer**: a relatively thick transparent layer of material intended to protect the sensitive layer of a storage medium from mechanical and chemical damage, and to keep dust out of focus; called a 'window' if spaced away from the sensitive layer and an 'overcoat' if in contact.

**pulse code modulation**: the method of coding a signal based on modulated peak values of pulses in electromagnetic or optical transmission of information.

**pulsed laser**: a laser which emits light in discrete pulses, as distinct from a continuous wave (cw) laser.

**pulse width modulation**: a method of data transmission in which information is represented by variation in the lengths of successive pulses in a nominally uniform train; abbreviated PWM.

**PWM**: abbreviation for pulse width modulation.

**radial**: along a line from the center to the outer edge of a disc.

**radial acceleration**: when a disc is rotating, the rate at which a track accelerates toward or away from the disc's center because it is not perfectly centered relative to the spindle hole or perfectly round.

**radial runout**: the peak-to-peak radial motion of a track relative to the rotation axis. The radial runout is approximately twice the concentricity.

**radio frequency sputtering**: a method of deposition of thin films of materials in which radio frequency energy is used to pull atoms or molecules of the material to be coated from the target onto the substrate; the process is done in a vacuum.

**RAM**: Random Access Memory device.

**read-only media**: optical media with a large amount of information replicated into the substrates. New data cannot be written onto read-only media.

**Red Book**: standards which define the characteristics of audio compact discs and drives.

**Reed-Solomon codes**: a class of extremely powerful ECC codes.

**recording threshold**: the smallest recording power that produces marks or pits.

**recording zone**: the annulus of an optical disc which is intended to support optical recording. The recording performance of an optical disc is guaranteed only within the recording zone.

**refractive index**: the ratio of the speed of light in vacuum by the speed of light in a material—a critical optical proerty, abbreviated n.

**replication**: the process of pressing quantities of compact discs.

**reverse bias**: the voltage applied to a semiconductor junction or diode with polarity such that little current flows.

**reversible media**: see erasable media.

**ROM**: Read Only Memory device.

**runout**: deviation from perfect motion of a moving object. For example, if a horizontal, spinning disk is slightly warped it will wobble up and down; this is axial runout. A disc with an off-center spindle hole will move back and forth relative to the read/write head; this motion is called radial runout.

**scanner**: a device which resolves a two dimensional object such as a business document, into a stream of data bits by raster scanning and quantization. Electronic scanning permits, capture, storage, retrieval and manipulation of data by computer systems.

**SCSI**: commonly pronounced 'scuzzy'; abbreviation for Small Computer Systems Interface. Same as SASI, which was changed to SCSI when it was adopted as a standard by the U.S. Government.

**seek time**: the time required for a head to move to a new track. The average seek time is defined for a motion equal to one-third full stroke, or one-third the width of the recording zone.

**semiconductor laser**: a term usually used to refer to injection lasers.

**sensitive layer**: the thin firm layer in an optical medium on which information is recorded; it may actually be composed of more than one layer. Materials encountered include many plastics, semiconductors and metals.

**sensitivity**: the laser power required for optimal recording at a particular scanning velocity with a particular optical stylus (wavelength, Strehl ratio, NA, etc.).

**servo**: short for servomechanism.

**servo cross-talk**: see 'Cross-talk'.

**servomechanism**: a combination of detector and actuator which continuously monitors and adjusts a variable signal as part of a control loop mechanism.

**settling time**: the time required for the tracking servo to lock on to a newly acquired track and eliminate transient tracking errors.

**shot noise**: electronic noise that arises because of the quantum nature of light and/or electrical current. Data channels are usually not dominated by shot noise unless the detector illumination and photocurrent are very small.

**signal-to-noise ratio**: see 'SNR'.

**SNR**: abbreviation for signal-to-noise ratio. One measure of the quality of a signal channel is the relative strength of the information signal and the background noise (i.e., everything in the channel which is not the desired information-bearing signal). In the common case where the noise is mostly due to a random process, it is sufficient to describe the noise by its power. This can be compared to the signal power; the result is the signal-to-noise ratio, measured in dB.

**solid-sandwich disc**: an optical disc formed by laminating two substrated together. The disc operates in a substrate-incident mode and is usually designed to be two-sided.

**stamper**: a disc created from a master used to replicate (stamp) working discs.

**stimulated emission**: emission of a photon that is stimulated by another photon of the same energy–the process that makes laser light (J.Hecht).

**Strehl ratio**: the central intensity of a focused spot divided by the intensity that would theoretically be present in the absence of all aberrations. The Strehl ratio is always less than 1.

**substrate**: the relatively thick part on which thin film media layers are coated.

**throughput**: computerese for the volume of work or information flowing through a system. Particularly meaningful in information storage and retrieval systems, in which throughput is measured in units such as 'accesses per hour' of data movement or retrieval.

**time base**: a frequency reference used to determine all rates of operation in a system.

**TIR**: 1.  total internal reflection. When light passing through a medium encounters an interface with a medium of lower refractive index, and if the angle of incidence of the light beam onto this interface is greater than the 'critical angle', all the light is reflected at the interface.

2.  total indicator of runout. The total amount of runout; see 'runout'.

**track**: a linear or circular path on which information is placed or stored.

**track jump**: the action of moving quickly from one track to another nearby.

**tracking servo**: the servo used in an optical drive to keep the reading and/or writing light spot centered on the information track, despite imperfections in the medium and drive mechanism as well as externally-improved shock and vibration.

**transducer**: a sensor which converts one form of energy into another.

**transfer rate**: the rate at which data is transferred to or from a device; especially the reading or writing rate of a storage device. Usually expressed in kilobits or megabits per second.

**transmissive read**: refers to an optical medium where the laser beam to be read passes through it, as distinct from 'reflective read'. Generally, though not always, the information-bearing surface of such a medium is a polymer-dye.

**transmissive**: allowing light to pass through, rather than reflect from.

**trilayer**: a remarkably successful multilayer coating structure used in writeable optical disc media to increase the amount of light actually absorbed by the layer to be heated.

**two-axis actuator**: a combination tracking and focus motor, that operates in a closed-loop servo system to position the objective lens precisely relative to the optical disc in both the axial and radial directions.

**unobstructed access**: the case in which the optical head may gain access to the recording zone of a disc from beyond the disc's periphery while the disc is on the spindle, without raising or lowering the optical lens assembly.

**vacuum deposition**: thin film application of material to a surface by particle transfer in a vacuum; common vacuum deposition techniques include evaporation, in which material is boiled off of a source with the vapor condensing on the piece to be coated, and sputtering; see 'radio frequency sputtering'.

**video disc**: a disc storing video information; it can be optical, as is LaserVision[tm], capacitance sensed as are CED[tm] and CHD[tm], or mechanical as is TED[tm].

**voice coil**: a type of linear electric motor, much like a solenoid. It consists of an electromagnet that encloses a movable element. As changing electric current is applied to the electromagnet, the element moves forward or backward at varying speeds, depending on the voltage of the current. Voice coils are found in loudspeakers, optical storage focus actuators, and radial head positioners for disc drives.

**WORM**: write-once-read-many: the broad class of optical recording media, on which data can be recorded but not freely altered. Data on write-once media can be altered by writing additional marks, but individual files cannot be rewritten.

**WREM**: Write-Read Erase Memory device.

**Yellow Book**: standards which define CD-ROM discs and drives.

# Index